SUNDERLAND INDUSTRIAL GIANT

SUNDERLAND INDUSTRIAL GIANT

Recollections of Working Life

MARIE GARDINER

For Gran and Granda, whose stories started it all.
And Mark, for the journey.

First published 2017

The History Press
97 St George's Place,
Cheltenham, Gloucestershire, GL50 3QB
www.thehistorypress.co.uk

British Library Cataloguing in Publication Data.
A catalogue record for this book is available from the British Library.

ISBN 978 0 7509 8120 0

Typesetting and origination by The History Press
Printed and bound by TJ Books Limited, Padstow, Cornwall

Contents

About the Author

MARIE GARDINER is a writer and photographer from Sunderland, now living in County Durham. After earning her degree in film and media, she worked as a broadcaster for a number of years, before starting a media company with her partner, which under the banner of Lonely Tower Film & Media, produces historical documentaries.

Acknowledgements

In writing this social history of Sunderland, so many people have offered their help, time and support. In a manner fitting to a long Oscars speech that you wish would end, I'd like to thank them here.

Firstly, to everyone who invited me into their homes, their workplaces, or met me on a freezing cold winter's morning to share a piece of your lives with me: thank you. I hope in telling your stories, I've done them justice.

Thank you too, to the people who made those all-important introductions: James Ramsbotham, Janet Robinson, Chris Hall, Michael Ganley, Sarah Stoner, Martin Dent and Neil Cuthbert. For organising times to interview in busy schedules: Ben Guy and Matthew Walker from Nissan; Lesley Callaghan and Nichola Kostyszyn on behalf of Sir Bob Murray, and SAFC.

Special thanks to Martin Routledge and Sunderland Museum and Winter Gardens, and to Norman Kirtlan of Sunderland Antiquarian Society for their continued help and support, and for letting me raid the archives with abandon. Image copyrights should be taken as a second acknowledgements list, and are credited individually.

As is the way now, much of our sharing of history is done through social media. To that end I'd like to say thank you to the Facebook groups who have been immensely supportive: Nostalgic Memories of Sunderland in Writing, Sunderland Tugs and Shipbuilding in Pictures, Vaux Brewery Collectables, Sunderland Antiquarian Society, and Sunderland's Heritage.

Thank you to Rob Langham for the sage advice, and to The History Press and my editor, Nicola Guy, for the opportunity to write this publication and meet some wonderful people in the process.

Foreword

Norman Kirtlan, Historian at Sunderland Antiquarian Society

As young children growing up in 1950s Sunderland, it seemed that giants played a huge part in our lives. My own giant, the one with whom I lived and called Dad, seemed to disappear into thin air every morning at seven. I would see him no more until late in the day, when he would return with his overalls and flat cap blackened with grease, his hands scrubbed red and stinking of Swarfega. It seemed that the smell of oil and grease was an ever-present companion to my Dad, except of course on Sundays, the working man's day of rest.

The daily ritual of returning fathers and home-cooked teas was played out in thousands of households in our town, and often, when the men came home, the women would be off themselves. A day's graft in the house and a night's graft in Ericcson's, Heppies or Jackson's ... there was always work for those who wanted it, and there were precious few who didn't leave school on a Friday and start their working lives on the Monday.

My Dad often took me on walks along the banks of the river, where more giants loomed large against the skyline. He would name the shipyards as we passed them: Bartram's, Thompson's, Austin's – and each would be alive with the sounds of toil and the crash of heavy hammers and rivet guns. Across the river stood the pit – it was never given its name, because he felt that it didn't need one. It was just the pit. It was here that his own father worked, deep in the bowels of the earth, blue scars etched into his skin from a hundred cuts and a patch that covered the eye he had lost to a kick from a pit pony. The pit would eventually claim his life, but that was just the way of the world. Men accepted that they would breathe in the dust from the mines and the fumes from the yards, but giants, they say, usually expected a sacrifice of sorts.

As I grew up and left my childhood behind, the town still seemed to be alive and energetic. Ships still filled the river, smoke filled the air and heavy industry offered opportunity to those entering a vibrant and demanding workforce for the first time. Much has changed since those years of plenty. One by one the giants of industry have died. The industrial cathedrals wherein men and women spent their working lives are no more, replaced by luxury housing or fresh green fields. All that is left are memories. And memories, like summer flowers, must be collected before they too wither and die. Marie Gardiner has gathered the thoughts, the feelings and the stories of men and women who lived among the giants. She has given their stories a precious platform to be heard. Her book, *Sunderland: Industrial Giant*, contains not only words, but voices that may never be heard again.

Introduction

The city of Sunderland, in North East England, is home to around 300,000 people. Although named for its main settlement, it also encompasses Washington, Houghton-le-Spring and Hetton-le-Hole, as well as a scattering of villages in the area. In 1992, the town was awarded city status.

As I was interviewing the people whose stories appear in this book, every one of them asked me if I was from, or, more endearingly, if I 'belonged' to Sunderland. It's a strange thing, the concept of belonging to a place. Even if we move away as students or young adults and spend most of our lives somewhere else, we refer to the place where we grew up, as *home*.

Having been born, lived and then worked in Sunderland for over thirty years, I've never really felt that my identity was strongly linked to the city and if that was the case, I wondered why, then, I'd chosen to write a book about it.

My family have been in Sunderland for a number of generations. During a recent trip to the battlefields of northern France, I by chance discovered a relative on the memorial wall at Arras. I'd decided to look up my surname in the book, and there, listed under Gardiner, was a David Gardiner from Hendon, Sunderland. I mentioned it when I got home and my gran told me that my grandfather's grandfather was killed near Arras. Some further research confirmed that they were one and the same, and we visited the site again the following year. It seemed like every time I had a conversation with my gran, now in her late 70s, I'd find out something new about Sunderland, her life, or our family, that I didn't know before. I wondered how many other families were experiencing the same thing and what the implications of that were.

The stories we hear from our grandparents are the ones we pass on, until eventually they wink out of existence, weakened by decades of retelling, changing details and forgotten names. It suddenly became really important for me to record some of these stories, not just the ones of my family but of others; of people who lived and worked in a Sunderland that seems long gone, but is really less than two generations ago.

The idea of identity is a strange and fluid thing. Sometimes our identity comes from how other people perceive us, and I think the characteristics of a town or city are no different. In early 2012, the *Daily Mail* published an article about how, according to government statistics, 'people from Sunderland were more likely to drink themselves to death than anywhere else in the country'. I read a story recently that pronounced us as the 'Facebook Capital of Britain', and a few days ago on Twitter I saw a photograph of Fawcett Street, taken by a local, of two Greggs bakeries side-by-side, with the pithy caption 'The Day Sunderland Peaked'. Don't ever say we don't have a sense of humour about our city. Still, these are not positive stories, and the saturation of this kind of reporting adds to how others are seeing Sunderland, and it's not in a good light.

During the day, even on a weekend, Sunderland can be an uncomfortable place to spend time. I was once working on a photography assignment in Mowbray Park when a man sidled out of the bushes like something from a bad comedy sketch,

Sunderland Museum and Winter Gardens. *Marie Gardiner*

to ask me what I was doing. He'd been sat in the shrubbery, drinking (and more, I suspect) and was slightly the worse for wear; this was lunchtime on a Saturday. That same day I spotted two people surreptitiously trading a package and money next to the Winter Gardens, about 5 feet from where children were playing and feeding the ducks. Mowbray Park is one of the oldest parks in the region, a grand old Victorian throwback which, restored in the 1990s and now kept in immaculate condition, should be the centrepiece for the city. Instead, it's somewhere you're told as a youngster not to walk through alone and never in the dark.

Sunderland is in a curious position at the moment, torn between its 'glory days', what might have been, and frustration at lack of opportunities – a feeling the city isn't moving forwards quickly enough. Part of this seems to be a struggle with identity. To the rest of the country, anyone with a North East accent is a Geordie (someone from Newcastle); there's a perception that Newcastle–Gateshead receives the majority of funding in the North East and of course, there's the bitter rivalry between the two football clubs. You accidentally call a Mackem (someone from Sunderland), a Geordie and you'll soon be put right.

Industry in Sunderland was something to shout about, even to the most recent generations. 'Did you know that we were the biggest shipbuilding town in the world?' is something I've been asked on numerous occasions. The high turnout of ships, coal or glass, has long been a source of pride: we worked hard, we made things that the rest of the country needed, ergo they needed us. When those industries started to decline and eventually disappeared, people found themselves out of work, and that sense of pride faltered.

Recently, during the uncomfortable period surrounding the UK referendum on whether to exit the European Union, Sunderland hit the headlines as the poster child for Brexit. We even made headlines in the *New Yorker*, 'Sunderland and the Brexit Tragedy'. John Cassidy reported:

> In the minds of many inhabitants of Sunderland and places like it, Brussels and Westminster represent the political face of an economic system that has ignored them. As Wake told de Freytas-Tamura, the Brexit referendum enabled disgruntled voters to 'poke the eye' of the political establishment. The tragedy is that this gesture wasn't just pointless – it was counterproductive. If the UK economy now enters a recession, which many economists believe is likely, Sunderland will suffer along with everywhere else. Unless the Brexit vote is somehow reversed, the residents of places like Sunderland will most likely be left to fly the Union Jack and fester.

The referendum has divided the nation, and Sunderland is no exception. The local newspaper, the *Sunderland Echo*, demanded an apology from the *New Yorker* for what they call a 'biased, patronising and grossly distorted picture' of the city. It's also played host to numerous arguments on its website and letters page; Facebook groups that have nothing at all to do with politics have suddenly erupted into vicious rows about who's right or wrong. The only time things really seem to be on a calm, even keel is when the rich history of Sunderland is being shared and talked about. Politics are long forgotten when Sunderland Antiquarian Society posts an old picture of Fawcett Street or when a photograph of the Wearmouth Bridge, packed with men heading to the shipyards, is shared and then exclaimed over with great nostalgia.

Our history and heritage bring us together; knowledge gaps cause us to ask questions and to seek answers. We need to be sure, however, that those answers are without the benefit of our rose-tinted glasses. This can only really come from the people who've lived here, had that vocation, and been through those experiences. Yes, our past was industrious, but it was hard, dirty and dangerous at times. Most importantly, though, it's our history, told through the people who helped carve that great reputation, that we still reminisce about. These are their stories, but in telling them, reading them and learning from them, they belong to all of us.

1

Coal Mining: Monkwearmouth Colliery

Coal mining was one of the region's first industries, and it became widespread in the North East around the thirteenth century. With good sea transport links near to abundant, shallow, coal seams it was the ideal place to mine. Jump forwards to the seventeenth century and Wearside was teeming with coal mines, commonly known as pits.

Mines covered County Durham's landscape, and headgear, engine houses and associated buildings dominated the skyline. Sunderland's largest mine, Monkwearmouth Colliery (or Wearmouth Colliery) opened in 1835 and was the last to still be in operation in County Durham's coalfield. Small, rural villages became colliery towns, and new towns sprang up. By the early twentieth century there were almost a quarter of a million coal miners working in the region, and more than half of those were in County Durham.

The North East saw a massive influx of workers, particularly in the later part of the nineteenth century, when people travelled to the North East for the high wages that coal mining and shipyard work had to offer. There's a perception that people of past generations didn't move around much, but for manual work and particularly for mining, where there was the prospect of a house and good wages, they would go where the work was. Conversely, when opportunities dried up during a depression in the late 1920s to 1930s, a scheme was put into place to help relocate those living in mining towns who no longer had work. Some 10,000 people left the North East as a result.

The culture of mining is heavily ingrained in the people of the North East, not only from the work itself but from traditions that have risen from it. The Durham

Durham Miners' Gala, 1983. *Sunderland Museum and Winter Gardens*

Miners' Gala (often locally referred to as 'The Big Meeting') is an enormous gathering, held in Durham each year on the second Saturday in July. The gala consists of a march through the city with a number of banners being carried to represent the various union branches, and often with the local brass band of a particular town or village accompanying the relevant banner.

The tradition of a gala started in 1871 and was born from the first union established a few years prior. In the twenty-first century, the gala is celebration of the region's rich mining history and on a good year it has been known to attract upwards of 300,000 attendees.

I meet Albert Holyoak in the Wearmouth Miners' Welfare, a small but well-cared-for social club just off Thompson Road in Southwick, where the ex-pitmen still meet regularly. I'm accidentally half an hour late, but Albert is good natured about it and asks if I got lost. The room is freezing cold, despite the fact that they've been hanging around for half an hour waiting for me; Albert explains that they're not allowed to turn on the heating. We chit-chat while I set up the equipment to record our interview, and I can already see Albert is full of character:

> We lived in Ross Street in Southwick. There was a big family of us, five lads and five lasses so there was plenty to do! When you've got five lads, you haven't got much copper. When your older brother grew out of his trousers, you got them because you were the next one down; nobody got anything new. I remember one Easter; my mother took me to the shop and I got a jersey and a tie. I was dressed like a lord! I'd never had anything new, everything was handed down in those days, couldn't afford not to. Hard times, but we got through it.

Albert left school at 14, as most young boys did at that time, and started work as an errand boy at Thompson's Red Stamp Store, part of a popular chain of grocery and kitchenware shops around the North East that would give out loyalty stamps with purchases. He worked there for a year before things were to change dramatically for him:

> One Saturday when I came home, my father said, 'You've got to put your notice in … the coal owners know I've got a son of working age and want to know why you're not at the colliery.' They told my father that he'd lose his job and the colliery house we were in if I didn't go into the pit.

This is something that occurred frequently due to the hold the coal owners had over the pitmen. If someone worked at a colliery, and transport to the area wasn't readily available, they were often provided with a house – the quality of which

An aerial view of Wearmouth Colliery, 1928. *Sunderland Museum and Winter Gardens*

would depend entirely on how prosperous (and how generous) the owner was. It's said that the mine owners of the North East took a more vested interest in their workers than elsewhere in the country. By comparison, wages in this region were generally higher than collieries in Yorkshire or Staffordshire, for example.

'If you had a father who worked at the colliery, you got a colliery house and you were tied to the coal owners. So that was it. I didn't want to go to the colliery,' Albert says. 'The coal owners put a gun to your father's head to make you go, otherwise you'd be sacked and lose your home.'

Albert recalls his first days at Wearmouth Colliery clearly:

> I went to colliery offices on a Sunday morning and the manager said I had to start work on Monday at six o'clock in the morning. They used to let the new workers stay on the surface for a few days, to get you used to it, but they sent me straight down the pit, I was surprised. The first time in that cage …

Albert sucks in his breath and shakes his head at the memory. 'The first day down the pit, they sent you down with an older lad and I was with him for about seven or eight days and then set up as a haulage hand.'

Lamp tokens (sometimes called checks or tallies) played a dual role. They let management know who was at work, but also ensured they knew who was still inside the mine if disaster stuck. Exactly how the token system worked could vary from pit to pit, but the majority issued one to each underground worker, who would exchange it for a lamp stamped with the corresponding number. When he finished for the day, the miner would hand in the lamp and receive his token.

We often retain the strangest fragments of information from our working life. One such piece for Albert is his token number, and it's something that is common among miners. They were told never to forget their number, and they never have:

> It was 1354. You had to get your lamp from the lamp room to go down the pit, and I was talking to the overman at the time and said there must be a lot of people come through here for me to be 1354. He told me that there were 3,500 people working at Wearmouth Colliery at that time. There were so many different parts of the colliery, you had Hutton, Harvey and Maudlin seams, and they all had to be employed. In those days they were all pony-driven putters, there was no mechanisation, it was hard slogging, you were ready for bed, I know that! You worked hard on the coal face, maybe filling as much as twenty tonnes of coal and you'd walk all the way out again.

The National Coal Board was created to run the coal-mining industry when it was nationalised, and started managing the collieries on 1 January 1947. It immediately began to invest money into the mines, making changes to increase and cheapen coal production by mechanising the process and investing in workers. 'We all got issued with overalls and trousers, kneepads, safety boots, where at one time you had to find your own,' Albert remembers, 'the Coal Board had its good points as well as its bad. They paid over a million and a half for new pithead baths. Before that, we used to come home covered in dust and have to bath in a big tin tub in front of the fire.'

Pithead baths made a marked difference to the life of a miner, particularly one forced to wash outside in the yard. Before the baths were available, coal miners would make the journey home filthy with coal dust and often damp with sweat and water from the pit. As such, they were often susceptible to chest conditions like bronchitis and pneumonia. The wife of a miner had her quality of life improved greatly by the introduction of the baths, too. Previously, as well as having the bath ready for her husband and the arduous task of washing his clothes, she'd also be battling to keep the house clean from a constant layer of coal dust. With the baths, miners could bathe and store their dirty work clothes in one locker and their regular clothes in another, so that they could go home clean.

Advances in technology and innovation helped to improve conditions in the mines, refining coal extraction, enhancing working conditions and even simplifying travel within the seams. Albert experienced much of this evolution first-hand: 'You had a 2-volt lamp and it was better than when you had an oil lamp. There were roadways under the ground and it was all whitewashed. I couldn't fault the machinery. When I first started, there were pit ponies and everyone had to walk.' Pit ponies were coming to the end of their time when Albert began working at the colliery, but their first recorded use was actually in the Durham coalfield in the mid-1700s and as such, they were a regular feature of mines in the area and seem to hold a special place in the hearts of many. The ponies were essential to coal

Miners leaving the cage, Wearmouth Colliery, 1966. *Sunderland Museum and Winter Gardens*

mining at the time, used to transport the coal and help haul it up to the surface. The treatment of pit ponies and the conditions they were subject to are hotly contested. Many say they were well-treated and healthy; other accounts tell a sadder story of injury, blindness and death.

Walking to work through the mine became problematic as the seams increased in length. 'When I first started at the colliery, it was a seven-and-a-half-hour shift,' Albert recalls. 'As the years went on, we were working twelve-hour shifts and fighting for shorter hours. There was so much travelling to do underground; you were ten miles out to sea.' By the time workers had got to their appointed place in the mine, they'd taken a considerable slice from their working hours. To combat this, man-riders were put in place. These were small trains designed to transport the pitmen and significantly reduce travelling time.

Working in mines was notoriously dangerous. Explosions, fires and collapses are stories all too common and the cemeteries of pit towns often contain grim reminders in the form of memorials and dedicated sections of cemeteries for people who lost their lives. Even though conditions improved over time and disasters and deaths became less frequent, particularly after the mines were nationalised, disasters still occurred in Albert's day.

One poignant reminder of this lies in the form of a memorial garden in Easington, just down the road from Sunderland, and the site of the former Easington Colliery. As Albert worked in Wearmouth Colliery on 29 May 1951, an explosion at change-of-shift time in Easington Colliery, rocked the area, trapping and eventually killing eighty-three men, including two rescuers.

Unless you've experienced it, it's hard to imagine a job where your life is potentially in danger on a daily basis. The days of firedamp explosions and collapses were, by now, rare and most deaths were due to falling down a shaft or being crushed by equipment. 'There were one or two fatal accidents at Wearmouth, while I was there,' Albert recalls:

> Shaft men used to examine the shaft two or three times a day, and they'd stand on top of the cage inspecting the brick work and making sure it was all safe. One day, two of them – they were brothers as well – fell down the shaft. The union used to have a Widows and Orphans Fund, and everyone donated a penny or tuppence, so the two lads in this case, and all the others who died, would have their funerals paid for. Sometimes the fathers who'd died had sons who still came to the pit. I remember them, and I see them now, grown up, with families of their own.

In the 1970s, coal consumption plummeted and the need for coal mining started to decline, with many pits attempting to reduce costs. In the early 1980s, Wearmouth

Colliery needed to cut its workforce and so they looked to their older pitmen to take voluntary redundancy. Albert decided it was time to go:

> I was 58 when I left the colliery and I said if I had six years on the grass after being down that black hole all those years … and I've had thirty-odd so far. I used to hate the pit but I had to stay because I had a family of four lads and I had a colliery house.

Strikes at collieries were not infrequent over the years, the earliest one on record being in 1765 when miners fought over having to obtain a leaving certificate before they could work at another mine. There was no union then, but the men banded together and withdrew labour for six weeks until they were successful and the certificate idea was scrapped. In the early 1830s, Northumberland and Durham miners formed a union, and in 1831, stayed out for a number of concessions, one of which involved reducing the hours worked underground to twelve a day rather than the previous eighteen. In 1844, the coal owners physically evicted families from their cottages using hired hands known as candymen, and hired labour from elsewhere to work the mines.

The miners' strike of 1984–85 is understandably a particularly difficult subject for the North East. The scars run deep, even after so long. Subsidising the mines was costing the government too much and unrest began after a statement by Ian McGregor, the head of the National Coal Board, expressed the Board's intention to close twenty collieries and lose around 20,000 jobs as a result. Miners downed tools on 12 March after the National Union of Mineworkers (NUM) declared it a national strike, despite the lack of a ballot. By the 14th, every miner in the Durham and Northumberland coalfields had joined the strike.

The strike was led by the NUM, represented most notably by Arthur Scargill, and was an attempt to prevent collieries from being closed. The aim was to replicate the success of a similar strike in 1972, where workers attained a victory by causing an energy shortage.

Although Albert had left Wearmouth Colliery in 1981, he stood by his fellow miners:

> We used to have a meeting practically every week, down in Monkweathmouth; Arthur Scargill was there sometimes. The best man in my opinion was Joe Gormley. He was the head of the National Union of Mine Workers, but in those days there was a ruling that when you reached 65 years of age you had to finish your job, so Joe had to retire, and Arthur Scargill took over.

Arthur Scargill signing autographs, 1981. *Sunderland Museum and Winter Gardens*

The long strike took its toll on the country, and Sunderland was no exception. 'People had families that had no wages coming in with the colliery being out on strike,' Albert remembers. The community rallied together to help those who had nothing, giving out food parcels, starting soup kitchens and holding collections. 'They used to have collection boxes and pails, they'd go in pubs and clubs and have boxes there. The miners were out nearly a year, it's a long time to be on strike.'

Bitterness and resentment flowed between families and friends who went on strike and picketed, and those who didn't. Many of these rifts would never heal. 'After the strike, the colliery villages fell apart, they all fell apart.' Albert looks stricken as he talks about it, as if it was yesterday rather than thirty years ago. 'It still goes on in some places, people don't speak to each other. Brothers who went back to work and brothers who were out on the picket line – a lot of families broke up in those days.'

The 'Davy Lamp' monument, in tribute to miners, outside of Black Cat House, the Stadium of Light, Sunderland. *Marie Gardiner*

The repercussions of the strike were far greater than anyone could have imagined. Pit towns snubbed the so-called 'scabs' who worked while others picketed. Nationally, the country was in disarray, with demonstrations often becoming violent. The *New Statesman* has this account of one particular night in 1984:

> I joined a group of NUM pickets who were avoiding the checkpoints in a clandestine cross-country yomp up Mynydd Merthyr, a wooded ridge above the village of Aberfan, where just two miners were going to work at Merthyr Vale colliery. The background to the mission was sombre because one of the two miners, David Williams, had been in a taxi a fortnight earlier when two strikers dropped a concrete post from a footbridge on to the car, killing the driver, David Wilkie. I remember how, early the next morning, the police charged on the pickets as the working miners' new taxi came into view. They trampled my photographer colleague Hugh Alexander to the ground, cracking several of his ribs.

This is a common story; old news footage shows miners who were still working being subjected to abuse, shouts of 'scab', and threats to their families hurled across picket lines. Protesters and police came to blows, windows were broken; it was utter chaos, and something for which my only personal frame of reference is the riots of 2011. Conceiving those lasting over the course of a year is unimaginable.

The miners returned to work on 5 March 1985, the strike unsuccessful; but instead of a fresh start, it was the beginning of the end. In 1987 the National Coal Board became the British Coal Corporation, and in 1994 it passed its final Coal Industry Act to transfer power to the Coal Authority. Assets were privatised and the British Coal Corporation came to an end on 26 January 1997. Many colliery houses were sold to landlords, which caused chaos in many small pit villages whose residents found themselves jobless and homeless within a matter of a few short years. Many of those looking for other work found that their actions during the strikes had resulted in a police record, which would prevent employers from hiring them.

The name Margaret Thatcher is enough to provoke an emotional and angry reaction, even today. 'Some say the best thing Maggie Thatcher did was close the pits,' says Albert:

> It wasn't a healthy job, you didn't get God's fresh air down there … but she destroyed the shipyards, destroyed the pits. These colliery villages, the community, they depended on the mines and when they closed, the roller shutters came down. Sunderland was a ghost; the shipyards closed, the engineering works closed, it

> put a lot of people out of work, but thankfully we've got Nissan, and I think it saved the day.

There's a sad end to my interview with Albert. His son, Terrance Holyoak, worked at Wearmouth Colliery for thiry-three years, following in his father's footsteps until its closure. Finding himself out of work, he got a job at Monkwearmouth School as a cleaner. On 27 August 2016, Terrance ended his life by jumping from one of Sunderland's bridges. Albert's son's death is so recent; you can see that talking about it is still very raw for him. 'They said it was mental illness,' he says sadly. 'How can it have been, when he worked down the pit for thirty-three years?'

Wearmouth Colliery closed in 1993, concluding its 158-year history, bringing to an end over 800 years of commercial coal mining in the region, and leaving hundreds of men without work. Today, the site is home to Sunderland AFC's Stadium of Light. The most prominent reminder of its great industrial history is a Davy Lamp monument; and the stadium's name itself was chosen in part as a tribute to the miners and football supporters, who emerged from 'that black hole', into the light, each day.

2

Shop Work: The Co-op and Joplings

Shops are funny things. I've worked in one, so retail is the one industry I can relate to on a personal level. Shops are a hub of social activity, even today when people are more inclined to order online and have it delivered to the door than they are to pop to their local store. You get to know people and their routines and habits: the presumably single man who comes in to buy all the microwave meals, the mother and son who pop in on the way back from the school run so he can buy a pack of whatever it is kids happen to be collecting at the time, the older folks who often use nipping out for the local paper as a good excuse to get out of the house and have some interaction.

Working in my local shop in Grindon, I always felt like an active participant in a – often comic – drama. We had big characters, like George, who used to come in and (in a light-hearted way) do his best to wind up the person currently serving at the till; the perpetually annoyed man who reeked of booze and always bought the same thing, a litre of cheap vodka and forty Mayfair cigarettes; the teenagers who used to come in regularly and steal the cans of lager stacked (stupidly) close to the front door. It's been a good ten years since I worked there, but I remember them well, and probably will for the rest of my life.

I'm invited to Moira Lawrence's home in Millfield and I'm surprised to see it filled with oriental paraphernalia, which Moira explains started when her daughter moved out and left a doll behind; the collection has continued to grow from there. Born just before the Second World War, in 1934, Moira is a middle child. 'I haven't moved far from the nest,' she laughs. 'I lived in St Mark's Road, down the road from where I am now. My father died and my mother was getting old and in ill health, so I moved to look after her.'

In 1939, Moira should have been attending the school on Chester Road, but the war had begun and soldiers were using the building, so Moira was forced to attend Barnes School. 'It was quite a walk for a 5-year-old, so my sister took me,' says Moira:

> I remember going to bed and having to get up because of air raids. We had a brick shelter in the yard and Dad had made bunks for us. He worked at the galvanisers during the day, making petrol cans for the war effort and then at night he was attached to the fire station, so a lot of the time it was my mam that had to get us all into the shelter. My gran used to come to our shelter because hers was always waterlogged, and she'd be carrying her little tin box with all her deeds to the house and everything in. Well, one night when we were staying at her house in Rainton Street there was an air raid warning, but the warden said she'd left it too late to leave and we'd have to stay. I remember the bomb going off and her pictures fell off the wall. The next morning, we went out to collect shrapnel.

I've heard about air raids and shelters from people before, of course, but one thing I'd not considered was what happened if you were at the cinema; a fairly popular pastime. 'We used to go to the pictures at Millfield,' remembers Moira, 'it would come on the screen if there was an air raid warning, and we'd have to leave for the shelter.' The cinema wasn't the only leisure pursuit affected, either: 'We couldn't go on the beach, because of the rolls of barbed wire and the fact they had soldiers stationed there.' When the soldiers eventually moved out of Chester Road, Moira was finally able to attend the appropriate school, which was much closer to home.

Before the Second World War, Britain imported a great deal of food, including the majority of its cheese, sugar, fruits and cereals. It also relied on imports for half of its meat supply, so it's no surprise that Germany chose to attack ships carrying supplies in an attempt to starve the nation. So came rationing: the Ministry of Food created a system whereby each person registered with a chosen shop and was given a ration book containing coupons. The shop would then be supplied with enough food for their customers and Britain could keep reasonable control over its provisions.

Moira remembers doing the shopping for her mother as a child:

The war had ended, but rationing was still in effect and it was still hard to get commodities. I used to go to the local Co-op store in St Mark's Road. You had to register your ration books with the shop, so they could receive the appropriate goods. The store wasn't owned by a person, by spending money you were part owner so you got dividends; I think was it 2/6 in the pound. They paid out twice a year, in the summer and at Christmas. For families in those days who were often quite hard up, it made a big difference. At Christmas it would buy a few extras

Hylton Road, looking east, 1930s. *Sunderland Museum and Winter Gardens*

> that you couldn't usually afford. I would have the children's [her siblings] books on me, so I could get oranges or bananas. You had to queue and they'd only allow you a few and then mark your book to say you'd had them so you couldn't come in again. One of the shops in Chester Road used to make gorgeous cakes, but they were limited because they couldn't get the flour, so we'd stand in the queue on a Saturday morning to get a cake; we'd get one for my mam and one for my nana.

'Producing co-operatively, to consume co-operatively', proudly proclaims the historical promotional video for the Co-op. And that's exactly what it was all about: customers would get the dividends that Moira remembers, in return for being members and shopping at the Co-op stores. Supermarkets, an American innovation from the 1930s, didn't hit the UK until 1948 and even then, it was a daunting new experience for housewives who were so used to the shop assistant measuring out the goods.

When Moira left school, she wanted to work in a shop, so she got a job aged 15 at the Hylton Road Co-op. 'It was quite a large shop,' she remembers:

> With a lot of staff. The manager, Mr Dodds, was small, like me. He'd say 'Are you sure you're 15?' I was the thinnest little thing you ever saw. One of the counters was quite high and you used to just see my head go past, which everyone found really funny so I got the nickname Little Moira.

With our pre-packaged goods and self-serve approach, it's hard to imagine how shops used to be. 'I was put on weighing and shown how you weigh a pound of sugar. We had hard blue bags that you had to put the scoop of sugar in, put on the scale and then take it all in properly, tuck the ends in and it had to be a neat little fold.' This makes me chuckle, as I can picture this perfectly. I have, on many occasions, opened a cupboard at my gran's, who used to work in a shop, only to find the flour and sugar neatly tucked up like this; apparently something that stays with you no matter how long it's been.

Moira continues:

> One counter was all dry goods like sugar and salt. Everything that would be in packets now was loose. The other counter had the bacon machine, butter and margarine, cheeses and everything like that. You served on the dry goods side, and they came in and gave you the ration books and then you collected their rations together. If they wanted anything else, like a tin of beans, then you went and got

that for them as well. When they'd finished you'd have to reckon it all up in your head, there was no cash machine or till, that's the way you were taught.

Soon after Moira started working at the Co-op, self-service shops were introduced in the North East, and Hylton Road was going to be the first of the Co-ops in Sunderland to make the change. Serving customers one at a time was less than economical and incredibly time consuming, particularly if there weren't many servers in the store.

Moira was moved to the shop in St Mark's Road, her local Co-op. 'It was very handy – I used to be out the house and over the road in five minutes,' she laughs:

I knew all the staff and the customers and the boss lived opposite, so if a boy brought me home he'd tell all the customers I had a new boyfriend! This shop was a smaller branch, on a corner site. It had wooden floors covered with sawdust, which we swept up every night and renewed every day. There were two counters, one for bacon and meats, and one for dry goods. The flour, we'd keep in big bins in the back of the shop and we used to keep the boxes of margarine there too, because it was packed in those days but lard and butter weren't. The butter came in a big barrel and they used to have to empty half of the barrel to carry it through to the front of the shop because it was so heavy. It took two people to lift it out, but not me because I was too small. And not the boss, he didn't lift anything [she laughs].

If anyone wanted cooked meat, we used to open big tins and they'd be cut by hand and wrapped in greaseproof paper. For one week for a person, I remember it was 5oz of bacon, 2oz of butter, 4oz of margarine, 2oz of lard, 4oz of cheese and one egg when there were eggs, which they're weren't always. The cheese was cut with a cheese wire, and if it snapped, it was repaired by knotting it and if you got your finger on the knot it would go in your fingers, so it made you very careful. The lard and butter used to need cutting and had to be exactly the right weight in one piece.

In a shop where everything was open and unwrapped, food hygiene as we know it today wasn't as much of a concern.

We had a covered-in yard which had just a standing tap on the wall, no sink or basin to wash your hands. The flour and the potatoes had to be taken up some stairs on a train and then dropped down a chute so that we could get them out of

Parnaby's Furniture Store, Hylton Road, 1970. *Sunderland Museum and Winter Gardens*

> the big bins at the bottom. Hygiene was out the window – there would often be a little mouse in the flour and I'd say to my mam, 'If they haven't got any packets, don't get any loose flour.' Our flour was very popular.

Once you'd been trained in one Co-op, you could work in any, as they all ran in much the same way. Moira was often sent to other branches to cover absences:

> I worked at Hendon, Southwick, Cleveland Road; everywhere! I loved it; I've always loved meeting people, I enjoyed serving so I was in my element. When I worked at the Southwick store, which was a self-serve shop, I found I really liked it and got on well with everyone, so they asked if I'd come over there permanently.

The end of rationing and the increase in popularity of self-service shops heralded big changes in the retail world:

> Everything was wrapped upstairs and then brought down. Customers picked up a basket and had to leave their bag at the cash desk as they weren't allowed to take bags into the store. People were really hard up, men were coming back from the forces and there were no jobs for them. The shipyards had been booming, but now they didn't need as many ships and so there were a lot of people out of work. We used to let people get goods during the week and pay on Friday night and then it started again the next week, so they were all living a week behind. Even more so at Southwick, it was a different area altogether. We used to write it all down and then transfer it all to a book in the evening.

It's hard to imagine in this day and age, being able to buy things on account without the aid of a store card with a huge percentage of added interest. Falling behind by even a week can be catastrophic to a family. The advent and popularity of the so-called payday loan – borrowing just enough to get you through a short period of time – will result in paying back several times as much, or worse still, starting a steep spiral of debt.

In 1954, Moira left the Co-op and went to work for Joplings on High Street West, in Sunderland's town centre. The store was established in 1804, trading as Jopling & Tuer – named after the two founders. It was sold to Hedley, Swan and Co. in the late 1890s, but they kept the name and so it became a household name for generations of Mackems and was often referred to as 'the Harrods of Sunderland'. According to local paper, *The Journal*, the store 'boasted the first escalator ever seen in the city, which at the time could only go up'.

Starting just before Christmas, Moira was enthralled by the display of toys, decorations and the Santa's Grotto. The Christmas spirit wasn't to last, unfortunately, as on 13 December 1954 Joplings caught fire. The *Sunderland Echo* reported:

> Narrow streets became valleys of heat, driving back the firefighters, with masonry collapsing on hose pipes. The worst moment came when firemen, playing their hoses on the front of the store, suddenly realised the building on the opposite side of the road had also taken fire. Within seconds, they were at the centre of a ring of fire; the intense heat could be felt by onlookers 80 yards away. By 2am, flames were tearing through the roof and the interior glowed like a furnace. The next day, all that remained was the shell of the building. It was Sunderland's biggest fire of the century.

These dramatic events saw the community pull together to help Joplings recover, setting up temporary areas around Sunderland to house their different departments. Moira remembers the temporary store that was built some six weeks later:

Joplings fire, High Street West, 1954. *Sunderland Museum and Winter Gardens*

Joplings fire, High Street West, 1954. *Sunderland Museum and Winter Gardens*

'They didn't have such a good display there, as it was so small. I remember snow was coming through the roof because it had been put up so quickly! Then they moved to John Street, which was bigger and a really lovely store.'

Joplings remained a firm favourite of Sunderland shoppers until May 2010, when the parent company at the time, Vergo, was placed into administration. Joplings closed in June of the same year and hopes of any last-minute rescue deals were dashed. As the building was part of Sunderland's architectural heritage and deeply woven into the fabric of Sunderland, grand plans were made to turn the former shop into a posh city-centre hotel. They fell through, and the building has remained empty, yet to find its permanent purpose. In January 2016, plans for a hotel and retail space were proposed, with the *Sunderland Echo* speculating that the site could be open to the public by summer 2017. In September 2016, it reported again to say the original firm had been outbid. As of March 2017, the Joplings building is still empty and it would seem plans for its use are no further forward.

Shop work and the community spirit that surrounded it clearly had a lasting effect on Moira. She'd wanted to work in a shop from when she was a little girl and then, as she'd moved from store to store, that hadn't waned. After having her daughter, her sister suggested they open up their own shop so that Moira could bring her daughter with her and still work. They opened a store in Whitehall Terrace, near the cemetery on Hylton Road, serving their community, for twenty-two years.

3

Vaux Breweries

Vaux Breweries had a rich and long history spanning almost 200 years. The company had a sign on the premises saying 'Established 1806' and though many sources site this as the beginning of Vaux, that isn't strictly true. Rather, 'Established 1806' refers to Cuthbert Vaux's first involvement with brewing; Vaux didn't form their own brewery until 1837 when C. Vaux and Sons was started on the corner of Matlock Street and Cumberland Street in Sunderland.

In 1844, they purchased a brewery in Union Street, moving again in 1875 when the North Eastern Railway Company needed the land to build Sunderland's Central Station. The brewery found a new home between Castle Street and Gillbridge Avenue, where it remained until its closure. The Vaux family ran the brewery until 1897, when Frank Nicholson joined as manager and secretary.

Sir Paul Nicholson, grandson of Frank, was knighted in 1993 for 'services to North East industry and the public'. He was managing director from 1971–76 and then chief executive and chairman from 1976–99, of Vaux Group (including pubs and later, the Swallow Hotels). He invites me to his home in County Durham to talk about his time with Vaux and the family history.

I'm not really sure what to expect when meeting Sir Paul; I've done my research, I've read his book *Brewer at Bay*, I've seen a couple of old television interviews of him being questioned about the brewery, I've even spoken to him on the phone prior to our interview, but I still feel nervous. We settle into a small den-like room in his large but inviting house and I'm somewhat comforted by the fact that Sir Paul looks slightly uneasy too. He starts by explaining how his family became involved with Vaux: 'My grandmother was a Miss Amy Vaux and my grandfather, whose business had been shipping in Sunderland, married her.'

Vaux Breweries, 1920s, Sunderland. *Sunderland Museum and Winter Gardens*

Frank Nicholson was promoted to managing director of the company in 1919. 'His brothers-in-law, who owned the company at the time, didn't want him to have any share in the brewery, but they let him run it because they were a little over-fond of the product,' Sir Paul laughs. 'The Nicholson family didn't own the brewery, and didn't even have shares in it until 1927 when the company merged with North Eastern Breweries and became Vaux and Associated Breweries. My grandfather was very dynamic and he became prominent in the brewing industry.'

Unfortunately, Frank started to suffer with his health, and in 1945 Sir Paul's father was forced to come back from the war early. When he died in 1952, his father took over permanently. 'When Father came back from the war, I used to go to the brewery fairly frequently and a lot of times I'd go around with the old head brewers, tasting the beers watered down with plenty of lemonade, so it was shandy,' he chuckles, 'so I knew all about the breweries for a long time.'

Despite his early induction into beer tasting, Sir Paul didn't actually train as a brewer. His father insisted he had a qualification before he came into the brewery, and so he trained as a chartered accountant. Once away at university, the attraction of staying in the south was powerful, but Sir Paul resisted:

> The call of the North East was strong. My father was keen that I should come back, so I did and I trained around the various bits of the brewery. I became

Draymen George Graham and George Stoker, 1958. *Pat Lowery*

> managing director in 1971 as my father unfortunately started suffering from dementia. It was a very difficult time, and I eventually took over as chairman and chief executive in 1976.

Heavy industry was beginning to decline in the 1960s, with mergers and closures in the shipyards in particular all too common. Unrest was in the air, and Vaux experienced a rare upheaval with its own staff in 1967. 'We had our troubles during the height of the industrial problems,' remembers Sir Paul:

> There was a 'work to rule' strategy, and the various anarchist movements had managed to insert some people who'd misled our workers to go slow. On New Year's Eve 1967, my father stood up on a wagon and dismissed all the people who'd taken part, which caused quite an outcry at the time. We negotiated, and we did manage to make peace and take back some of the workers. It was a period of great uncertainty. On the whole, though, we did have very good industrial relations in Sunderland; we were unusual in that respect, and we were widely respected, I think. People who worked for Vaux, many of them families just like my own family, really believed in the ethics of the company, which was giving good service and looking after people.

Many companies like to wheel out the company values, but for many it can be a box-ticking exercise; pretence of care. It was one thing to hear of the principles of a company from someone at an executive level, but another to hear it from a former employee. I meet Ed Forster on the wasteland where Vaux once stood. He looks around, clearly picturing where the buildings used to be.

'The warehouse was over there,' he says, gesturing:

> The guys in the warehouse used to joke on because we were all great Sunderland football fans, so there was always good chat about the football and about the brewery. Overall, during the time I was with them I found them to be absolutely fantastic. The benefits that you got in kind worked out very well for my two sons, who were at university. You used to get a pack and a half of beer every month, which was thirty-six cans, and regularly they were transported to Birmingham and Swansea University; the lads at the uni used to wait for me to arrive with it. You could buy beer at a good rate too.

'You got shares in the company and you could buy into a savings scheme which allowed you to buy more shares,' Ed explains:

With the sad demise of the brewery, you got a premium rate for those shares, so if you'd kept them then it worked out very well. All the Vaux pubs had a good name and so did the beers, they were massive. Hundreds of people were employed, not just by the brewery but in the pub trade and the hotel trade as well. One thing that always sticks in my mind with Vaux, is the managing director, Frank Nicholson [Sir Paul Nicholson's younger brother]. He was great with everyone, he knew everybody by their first name, and knew what was going on. So much so, that when my son started working in the brewery, Frank knew all about him and spoke to me about him.

If you were to ask those in Sunderland who are old enough to remember Vaux, the memory that stands out above everything else, it would be the horses. My gran's house is on a major route in Sunderland, and as a child I'd often hear the tell-tale clip-clop-clip-clop noise of the Vaux horses and run to the window to catch a glimpse. They were an iconic part of Sunderland and a mainstay of the brewery.

Sir Paul is very much an equine enthusiast. As well as having ridden horses all his life, he once also competed in the Grand National, and twice won the Fox Hunters'.

Vaux horses crossing the Wearmouth Bridge, 1985. *Kevin Lane*

This really surprises me, as Sir Paul is a tall man and he explains the rigorous dieting needed to get him down to the right weight. It makes sense then, with Sir Paul's family being horse lovers, that they would find a way to keep them as part of the business:

> The nice thing about Vaux was the horses: they were a feature of Sunderland, taking the beer around. It is rather sad that I read in the *Sunderland Echo* recently about the death of the last Vaux horse, a horse that I knew and used. I managed to get some of them transferred to Beamish Museum, where they were very much a visitor attraction. It was an economical way of distributing beer over a short distance and it raised the image of the company. It was an effective way of getting publicity relatively cheap and they were very much part of the company.

Most people have family snaps up on their walls; Sir Paul has family paintings, and he very graciously shows me his horse ones, including some of the Vaux horses, enthusing about the circumstances surrounding them. There are pictures dotted around on tables of Sir Paul with various members of the royal family, he was formerly the Lord Lieutenant of County Durham. It would be quite intimidating, if not for the fact that he's completely warm and self-effacing all the while.

Ed, too, remembers the horses well:

> When I went into the brewery, I spent a lot of time in the stables. I used to love going in there, and used to get taken around to see the horses and the maintenance of the tack that the young lads used to do, I was given a lot of information about that. Horses were a symbol of breweries and they used to go around town delivering the beer, while the wagons went on the further calls. Everybody loved to see the horses.

After interviewing Ed, I receive an email from Sir Paul's brother, Frank Nicholson, asking if he can be of any help, and we agree to meet on the old Vaux site. He pulls alongside me in his car to say 'hi', and I know immediately that I'm going to like him. Frank is full of energy, bouncing on his toes, smiling often, maintaining eye contact and using my name frequently. I can already see why he has such a reputation for engaging with his workforce; he makes a real effort.

Frank was managing director of the brewing and pub division of Vaux Group for its last fifteen years of business. He's good-humoured and modest about his induction into the company: 'It was by virtue of nepotism; being the chairman's youngest brother,' he laughs. 'We were only ever hired hands. Frank Nicholson, my grandfather, arranged the merger of C. Vaux and Sons with the other big Sunderland brewery called North Eastern Breweries, on Hylton Road, who were in horrible trouble.'

I think most of us can remember the first day we started working somewhere, and for an iconic company like Vaux, it's no surprise that Frank can vividly recall his first day in 1981:

> I knew quite a few people at the brewery. George Pickford, who'd been my father's chauffeur for years, was still working there. I arrived and I didn't know where to park, so I parked in the staff car park and George came out and said 'What are you doing parking there?' I said, 'Well, I'm an employee now,' and he replied, 'Yes, but you're a different sort of employee, I'll take your car.' It was extremely embarrassing, because I didn't want attention drawn to me; I was very shy and retiring then. I walked from the chauffeurs' hut to reception where I was greeted and then shown to my office. I had an office door with no windows in, it was still pretty formal in those days, and a face appeared around the door saying, 'Hello Frank, I'm Jack,' and it was Jack Clark, the northern regional manager for the pubs. He was the first character I met at Vaux and we've remained great friends ever since.

Frank was keen to learn from the ground up and having met him, I can fully believe he's the kind of person who wouldn't want to take a privileged position in a large company for granted. 'I spent time in different departments and quite a while working in the Grindon Mill,' Frank recalls. 'I know the Mill more intimately than any other Vaux pub. Dave and Sandra Spinks were the managers then. When people say 'you wouldn't know how to run a pub,' that's not quite true!'

Frank asks me if I know of the Grindon Mill and I explain that it was the local pub near to where I grew up, and I can remember my granddad going for a pint there when I was a child. When I got the bus home from town, it was the Grindon Mill stop that I'd ask for. It started to struggle some years ago, when drinking at home became more popular (and cheaper) and gastro-pubs were developed. This change heralded the death for many rough-and-ready local pubs and unfortunately the Grindon Mill was one of those. The building still exists, but today it's a gym; the mock-Tudor façade ripped off in favour of some questionable wood planking.

I wonder which stop the kids ask for on the bus today; the Grindon Gym doesn't have quite the same ring to it.

As Frank talks, he starts to tail off, glancing over my right shoulder towards the car park. 'Oh, can you see that, I think I'm just about to get a ticket.' I turn around

The Grindon Mill, Grindon, Sunderland, 2017. *Marie Gardiner*

to see a traffic warden in front of Frank's apparently ticketless car, writing down the details. I encourage him to pop across and see if there's still time for him to stop her and buy a ticket. He leaves, still wearing his microphone, so I'm able to hear as he apologises profusely to the bemused warden and effortlessly charms her into not only scrapping the citation she's writing up, but also into helping him use the ticket machine. He ends up with a small group around him, all eager to assist, pulled in by his infectious good nature. Polite to a fault, he thanks them, says his goodbyes and trots back over to resume our interview; fine averted.

'When I was at the Grindon Mill, we had tank beer,' he continues:

> Brewers talk about barrels of beer which are thirty-six gallons, and tank beer was five barrels. The tankers used to set off first thing from the brewery, about six or seven in the morning and they would usually do the outer districts first and then Sunderland later, but since I was in the Grindon Mill they thought, 'we'll get young Frank out of bed early'.

He laughs at the memory and shakes his fist in mock outrage. I get the feeling that Vaux workers felt they could have some fun with Frank, knowing he'd take it in good humour and something they'd rarely get away with, with someone in an executive position in a large company. It's part of what's earned him his reputation as a good boss, a people person.

Whenever Frank is mentioned in an old newspaper piece on Vaux, it's always stated that he knew every worker by name; that he could ask after the families of his workers, and was quick to praise and reward. When I mention that Ed speaks incredibly highly of him, he brushes the compliment off, looking down and smiling shyly. He immediately flips the conversation around to talk about how great the Vaux employees were:

> There was such a variety of people who worked here, you had skilled tradesmen, a French polisher in the workshops, and carpenters and joiners doing a lot of work in the pubs and hotels. Then you had customer service people who looked after the pubs, people who looked after the installation of very sophisticated equipment, engineers in the brewery who had to keep the plant going ... it was a very hi-tech operation. It not only had a modern brewery, but it had modern packaging facilities: canning lines, bottling lines, plastic bottling lines; these were pretty sophisticated bits of equipment, so you had the technicians and then the people who actually worked on them and they were equally skilful.

It's generally a smart move to be nice to your employees; if you like someone then you want to please them and do the best job you can, and I've often wondered why more managers don't seem to understand this. As well as it being advantageously sensible, I get the impression that Frank genuinely did, and still does, care about the Vaux staff:

> People tend to forget now, the variety of characters that were there. A big mistake of management is they never ask what people do outside of work or about their families, but I was always very interested to learn what people did outside work and it amazed me. Someone who'd worked in the warehouse for thirty years would be a scout leader, leading a troop of fifty scouts at weekends. I remember one young man in the warehouse, I asked him what he did and he said he was in the navy, in submarines, working with weapons and was responsible for ensuring the nuclear missiles were ready to fire if need be. I said 'Why on earth aren't you still doing that?' and he replied, 'I wanted to come back to Sunderland to be with my family and I'm happy to be here working in the warehouse, I enjoy it.' If I

> hadn't engaged him in conversation, I'd never have known that we had someone who single-handedly could've caused the next nuclear war, not that I said that to him!

'That was always the fascinating thing for me, the variety of character we had in the workforce,' Frank smiles:

> Every workforce probably does if you delved into their backgrounds, but particularly in Sunderland, they're very friendly. At Vaux, if you treated people decently, they treated you well too. We had great industrial relations on the back of nice people being dealt with in a good, kind way. I would get annoyed with senior managers because the worst thing they could do would be to walk past someone without greeting them; or say 'Hey, you'; that was a sin punishable by death if anyone said 'Hey, you'. It's 'How are you?' or 'Thank you very much for what you've done.'

Frank and Sir Paul certainly look like brothers, but have very different personalities. Sir Paul is warm but reserved, whereas Frank is outgoing, and I imagine that each was perfect for the role they were expected to fill. I've read a report describing Sir Paul as 'cool and distant', but I don't see that side of him at all, as he takes out quiche and salad from the kitchen and I help him to dish it out – he's the perfect host, even providing lunch. He takes an interest as we eat, asking questions and telling me about his own experiences.

When Frank talks about his brother, he does so with genuine admiration and affection, 'Paul is more distant in the sense that he was the chairman and he was *Sir* Paul,' he says, leaning forwards and emphasising the *Sir*:

> The people, including myself, had huge respect for him. His role wasn't to be so close to the people, but mine was and I tried over my years to get the senior managers to think the same way, that you have to treat people decently, you have to remember to ask them how their families are, you have to remember to say thank you and it's easy to say but it's amazing how many places it's not put into practice.

When you hear about Vaux now, how popular it was, the jobs it brought to Sunderland and even how successful it was in business terms, it's hard to understand how things could have gone so wrong. 'The problems started when, approaching 60, I decided I wanted to cease being chief executive,' says Sir Paul, 'and sadly, I didn't plan my succession carefully enough.' At the time, Sir Paul was a strong advocate of keeping the business interests together:

> But in hindsight, it would have been better if one had split the company at that stage into the breweries and hotels. The corporate finance advisor advised against that and that was the start of the trouble which culminated in the closure of the brewery in 1999. From the shareholders' point of view, the share price was something like forty times what it had been when I first took over so the shareholders did fine but the people who lost their jobs didn't.

Doubt has been cast over whether the brewery would have still been a success today, had it continued. At the time of its closure, it was a profitable enterprise and Sir Paul believes Vaux would have simply moved with the times:

> The brewing industry has changed enormously. It was destroyed by the Thatcher government, not that I blame Thatcher, but they recommended that the brewers shouldn't own more than a certain number of pubs and that made it uneconomic for a lot of the bigger brewers to operate, so they sold up. There were six major brewery companies and we were one of three in the second division. Now there isn't one. Microbreweries are now very popular and there's a very big industry growing around micro-gin.

Microbreweries are breweries that produce mostly small amounts of beer (or spirits like gin, whisky and vodka) and tend to be independently owned. Rather than concentrating on volume of production, there's usually an emphasis on how the product is brewed: experimenting with different flavours, for example. Some microbreweries – or craft breweries, as they're sometimes known – have created products so popular that they've then been bought out by bigger corporations. We can't know, but it's possible that Vaux would have made the natural evolution into this 'craft' style.

Frank describes how he and Sir Paul tried desperately to save Vaux:

> Our board fell out with itself and we committed corporate suicide. My brother and I were in favour of keeping the breweries and pubs, and running the whole thing as an integrated show, but nine of the board said no we want rid, and to be an hotelier, not a brewer and publican. So the argument raged for a short time before the decision was made to get out of brewing, which left me absolutely high and dry. I tried to do a management buyout of the brewing division, which

> started on 18 June 1998. I remember it vividly, it was the board meeting that said we're going to get out of brewing, and 19 June was me saying I'm going to try and buy the breweries and at least some of the pubs. For the next six months we slaved away to construct a deal, which at a very dramatic AGM at the end of January 1999, saw us agree terms, subject to that awful phrase, 'due diligence'. We got the funding, we had to raise £75 million.

He adds as an aside, 'which is quite a lot of money,' something I definitely don't need to be told.

When you're talking about figures into the millions of pounds like that, it's impossible to imagine it. Smaller amounts we might equate to the price of a car, or a house for example, but what can we compare £75 million to? To give the figure some context, the Stadium of Light was built in the mid-1990s and an extension added in 2000, bringing the total figure to £23 million; less than half of what it would cost to buy out Vaux and most of its associated business interests, some 300 pubs.

Frank continues:

> It was a very dramatic day because everyone thought we were all safe, but in the end the due diligence that we were required to do fell far short of what the group thought we should be doing, and they pulled the rug from under us. Our brokers said to the board, 'If you've got divided advisors, you've got to close the brewery'.

In March 1999, the board announced that it was going to close Vaux. 'Thereafter started a horrible period of winding things up,' Frank remembers. 'The only thing that kept one going is that we were all in the same boat together. I was made redundant along with 650 other people.' If I'd heard that from any other top-level management, I think I'd have rolled my eyes, but with Frank, I believe it. He's spoken so enthusiastically of the business, starting from the ground up to learn absolutely everything he could and gushing about the people he worked with, it's clear how emotionally invested in it all he was, and to have faced it with everyone else, people he considered friends as well as colleagues, must have been some small comfort:

> There's an element of when your back's to the wall, if there are enough of you with your backs to the wall it's not quite so bad. It was a terrible time and I'll never forget it. I don't think there's a day goes by that I don't think about what might have been, and I think a lot of people feel the same.

Vaux were very much at the heart of the community in Sunderland. As well as being involved in a variety of local enterprises, they also sponsored Sunderland AFC's football shirt and are, in fact, the longest-running sponsor the club have had to date, with shirts proudly bearing the Vaux or beer logo from 1985 until 1999. Such was their relationship with the club, that later, when the brewery faced closure, the *Sunderland Echo* handed out red cards so that fans could protest. The *Echo*'s report read at the time:

> The red card protest came ten minutes into Sunderland's 3–0 Saturday win against West Bromwich Albion. Swallow Group has rejected a second buy-out bid from the management of Vaux Breweries. The red card protest against the potential loss of 650 brewery jobs was organised by The Wearside Roar fanzine who also organised a petition to save the brewery. Editor Tom Lynn said fans were powerless to protect the brewery's future. He added: 'Visually it was very effective and basically showed the strength of feeling at the potential closure. There's not a lot more that the average person on the street can do now.' The silent protest had double significance for Sunderland fans, who learned that the Vaux logo – which has featured on the clubs shirt for more than a decade – will be replaced by the Swallow Hotels' crest from May.

For the city of Sunderland to see Vaux go when it was still reeling from the loss of its shipyards and, more recently, its mines, must have been heartbreaking. Having worked for many of the heavy industries in Sunderland, Ed remembers how he felt at the time of Vaux's closure:

> You've had the brewery go, the shipyards go; engineering, the pits … I don't know whether I'm a jinx on them, I worked for most of them at one time. Not one of those places is here now. I look at the town and I feel so sad to see what's happened. The economies did go, and that's why there's worry about Nissan, as it's the only big employer left.

During our chat, Ed had mentioned speaking to Frank after the brewery closed and asked him if he'd seen the site since; he'd replied that he couldn't bear to drive past it as he felt so sad about its loss. I was keen to ask Frank if he'd really stayed away, or if it was like a scab that you just can't stop picking. 'I never came anywhere near the site,' he states emphatically:

> I used to go to the Stadium of Light regularly, but I'd go via Hendon because I didn't want to come past the brewery. I can't think of driving past the empty

Vaux brownfield site, 2017. *Marie Gardiner*

> buildings once, I'd find any other way not to do it. Of course for the people who lived and worked in Sunderland, they were reminded every time they went by it. It left a wound, which has left a scar, very physical and emotional, in many peoples' lives.

He gestures around the still-empty site. 'The trouble with a scar is that unless you graft new skin on to it, it remains a scar. It's now coming up to eighteen years – the buildings stayed for a couple of years and then were cleared – and it's still bare other than a car park and a new road.'

I still have memories of the Vaux site, despite being very young when it was closed. I'm sure it was a big ugly industrial building, but there was something about coming around the corner and seeing the blue walkway with the gold letters and the chimney with Vaux written down it, that just made you think *home*. It's weird to have that association with something so industrial, but I can imagine similar feelings were felt around the shipyard cranes or the headgear of the mines. It's unattractive, it's dirty, but it's ours. Later, seeing the buildings chewed down – and they were chewed down, there was no glamorous, implosive demolition here – and seeing the skeleton of that same walkway, the covering and signage stripped away to bare wood and metal; was heart-wrenching. There was no quick and painless ripping off of the sticking plaster for Sunderland.

The continuing disappointment with Vaux ceasing is that the site has lingered on as a wasteland for so long, a constant reminder of the loss. Various deals and plans have come and gone, but very little has actually happened. Sir Paul remembers that when the brewery was finally pulled down, 'there was all sorts of talk about development and very little happened for the next fifteen or sixteen years, it was basically a desert. If one thinks that fifteen years of employment was lost, there'd have been about 600 people dependant on that site, and the man years would be something like 10,000.'

For many, Vaux was one of the last, big industrial employers in the city and with its closure, eyes turned to car-making firm Nissan to keep Sunderland going. 'Nissan was the best thing to happen to Sunderland for a long time,' says Sir Paul. 'Sunderland was the biggest shipbuilding town in the world and with that going and then the mining industry, it created huge problems. Fortunately, Nissan came and filled an awful lot of the gap. Vaux going was rather later than the others but it certainly was a tragedy for Sunderland.'

Frank agrees that Nissan is incredibly important to Sunderland, but wishes it was a home-grown business:

> Delighted as I am that Nissan is here, it's not an indigenous company – they arrived in the mid-80s, so they brought prosperity to the edge of Sunderland. Before their arrival, the early '80s saw the declining of the shipyards. The last shipyard closing in '89 coincided with the launch of the regeneration initiative, the Wearside Opportunity [of which Frank was chairman], which promoted Sunderland as an advanced manufacturing centre. Leading figures in the community came together to agree on a vision of the way forward and then we were supposed to get on and do some stuff, to prove it was achievable. We cleaned up some of the industrial areas, we got the Training and Enterprise Council, and we got 45 million pounds – with the benefit of hindsight, I'm not sure what that was spent on! We tried to re-establish Sunderland as *the* centre of manufacturing.

While Sunderland never really became a centre for manufacturing, it has started to pick itself up off the ground and dust off its knees. Frank muses, 'You talk to people now who don't even remember Vaux. You ask young people if they know of it and they'll say "I think my dad's friend worked there maybe." So memories do fade, it's rather characteristic of towns that have been in decline and are coming back, if Sunderland is coming back.'

The Vaux site is finally being developed. As I stand talking to Frank, a few men in hi-vis vests wander across the wasteland, pointing, deep in discussion. Plans are now in place to build offices, retail and leisure areas. To continue and complement The Keel Line art installation – a tribute to those who worked in the shipyards and designed to be the length of the *Naess Crusader*, the largest ship ever launched on the Wear – is a planned sculpture based on a shipyard crane. It's the start of something, and it's something the people of Sunderland can get behind and enthuse about. Whether it's the beginning of Sunderland *coming back*, though, remains to be seen.

4

Glass-Making and Pyrex

In AD 674, Benedict Biscop employed glaziers from France and Rome to make stained-glass windows for the Monkwearmouth monastery. It was the first record of glass being crafted in Britain and the beginning of a long relationship with glass production.

Making glass was time-consuming and expensive, so there was a considerable gap before glass-making started on Wearside in earnest. Leap forwards to the 1690s and the first glasshouse, Sunderland Company of Glass Makers, opened in Deptford; however, production was still limited due to expense. Mostly practical items were made, like windows and bottles, which were exported.

By the 1800s, Sunderland had seven bottleworks and three glassworks, the largest of which was James Hartley's Wear Flint Glass Works; the biggest in the country and responsible for manufacturing a large proportion of the glass used in the construction of the Crystal Palace. They are also said to have produced a third of all the sheet glass in England. By the mid-1800s there were a staggering sixteen bottleworks in Sunderland, churning out upwards of 60,000 bottles per day.

When glass pressing was invented, it allowed glass to be produced in a much quicker and more cost-effective way. Molten glass would be pressed into a mould, eliminating the need for every piece to be individually blown and so allowing for mass production.

J.A. Jobling and Co. Ltd, formerly Wear Flint Glass Works (they'd dropped the 'flint' to better represent what they were making) were one of two major, pressed glass firms in Sunderland. After the First World War, the days of middle-class women having domestic staff began to fizzle out and, for the first time, many women were forced to find their own way around the kitchen.

Inspecting and packing Pyrex lab ware. Wear Glass Works, 1922. *Sunderland Museum and Winter Gardens*

Jobling's started making art-deco glassware known as Pyrex in 1922, after the company began to struggle and Ernest Jobling Purser (nephew of James A. Jobling) decided to buy the rights to test a new technique he'd heard about in America, for making glass that would be oven-proof. In fact, Pyrex had been produced by US company Corning Glass Works for at least a decade by this time, but for Sunderland, it reinvigorated the glass industry.

Pyrex became the must-have item for housewives and one of the first products to be marketed directly to them. A dish you could put into the oven and take straight to the table was so advantageous that the use of it was further encouraged during the Second World War. Pyrex jumped on rationing shortages and urged women not to waste food. Instead, they should prepare casseroles in their Pyrex dish, they'd save fuel because it retained its heat and they'd even save on the washing up!

Jobling's were able to secure a license to sell Pyrex worldwide, and casserole dishes, dinner services and bowls were turned out of the Millfield factory in their thousands. Shares in the various interests were bought and sold frequently at this time and in 1954, Corning bought a 40 per cent share of the company. In 1958, Jobling acquired Quickfit and Quartz, who made laboratory glassware.

I meet Brenda Forster (née Calvert) via her husband Ed, one of our Vaux workers. They invite me round to their light and pleasant house in Herrington. Brenda makes coffee and insists she probably won't have anything at all of interest to tell me. I'm fairly convinced she will; it's often the people who think they have nothing to contribute who have the most fascinating stories.

She perches on the edge of the sofa; a slender, pretty woman who I later find out (much to my disbelief) is in her 70s, and tells me about growing up in Sunderland:

> I was born in a cottage in Williams Street, where we had no hot water. I don't remember the war, but I remember that my mam used to put me under the table when the bombs were dropping and a bit further down the road, my uncle lived

Grinding domestic glassware. Wear Glass Works, 1922. *Sunderland Museum and Winter Gardens*

> in a cottage that was bombed. We moved up to Scruton Avenue in Humbledon when I was 3. We had hot water and an inside loo; it was just inside the back door in a porch, where before we'd had to go down in the yard.

As with many families whose men worked at the shipyards or for the coal mines, Brenda's father worked at Lambton Staithes, for what was then the National Coal Board. 'It was near where the Stadium of Light is now, before you go over the bridge, down on the river,' she explains. 'The coal used to go to the ships on the riverside and the people who worked for National Coal used to get free coal, so a load used to be dropped at the gate. Dad used to come home filthy.'

Brenda went to Barnes School, leaving in 1958 at the age of 15:

> I went to night school, where the Civic Centre is now. It was West Park School in those days and I did shorthand, typing and English. You had to write to different places to ask for work, and I got a letter to say to go to the Wear Glass Works for an interview with Miss White. She said straight away there was a place for a junior clerk at the Pipeline factory. I loved her, she was a lovely personnel lady, and was really great all the time I worked there.

'I had to go into the factory to take a test and when I passed, I went to the Pipeline factory the next day to meet the manager, Mr Hindmarch,' Brenda remembers:

> I was very shy. I tried the door and it was locked, so I went home. The next morning my dad took me back down and made sure I saw Mr Hindmarch, and it turned out that the receptionist had locked the door while she went to the ladies, so it cost me an extra day to get my interview!

Thankfully, Brenda's shyness didn't cost her the job, and after meeting Mr Hindmarch and being shown around, she started work at Pipeline on Pallion Trading Estate:

> I worked in a general office where there were about twelve of us, so as a junior, I did all the running around doing errands and making the tea. As I got older and the girls were leaving to have families, I got a promotion to be in my own office at reception, and did typing and the switchboard and things like that. It was a really busy trading estate, it was hectic at dinner time. I got an hour and a half for lunch and I'd have to run down Pallion Road to get the bus to go home for lunch, and the Hepworth's girls and the shipyards were all in Pallion Road, it was hectic; it was a lively place in those days.

St Luke's Terrace / Pallion Road, 1971. *Sunderland Museum and Winter Gardens*

St Luke's Terrace leads down from Pallion Road, but often the whole stretch is referred to as Pallion Road, and it can still be described as thriving today. Driving down it is a heart-in-your mouth crawl, as you navigate the speed-bumps, zebra crossings and parked cars that dominate the busy street. Some independent shops still survive, but they're dwindling, and mostly it's now made up of takeaways – one small row of shops consists of an Indian takeaway, Chinese takeaway, two pizzerias and a fried chicken shop – and no fewer than three bookmakers. The local butchers and fruit and vegetable sellers are still hanging on, but with so many cheap junk food options available, you have to wonder for how long.

Workplaces were still heavily dominated by men in the late 1950s to early 1960s, but it was becoming more common for women to hold a job, and they had started to branch out into occupations than other admin. My grandmother worked at Pyrex for a time and part of her job was to put stencilled image transfers on to dishes. Brenda recalls that there were also a lot of women who worked in packing:

> There were a lot of men in the factory, but the ladies packed things like big flasks which had to be put in boxes with straw to keep them safe. On a Friday, a big maroon train container used to come and all of the scientific glass that had been done over the week and had been checked and packed, went into this big, long container. I had to type up all the instruments in each box and the invoice had to go into the cases, so it was always very busy on a Friday. Then the railway man would come and take the container away and it would go down to Stoke-on-Trent and then was distributed to different places from there. There was a director, Mr Gill, he was really lovely to me. When I was on reception he used to come in and talk to me. Some of the managers were snobby but he was always lovely; we were treated well generally. The girls got less money but I don't think it was a sore point where they were fighting for more pay, or wanting unions to stick up for them.

St Luke's Terrace looking up to Pallion Road, 2017. *Marie Gardiner*

I'd always imagined the glass factories to be rather tame compared to the industries down the road like the shipyards and mines, but Brenda remembers the factory being loud and dirty: 'It was always dusty, everything you touched was covered in glass dust. It was really noisy with the machines always going and if you went to Wear Glassworks, you used to see the plates, cups and casserole dishes coming off and the furnaces ...' she pulls a face at the memory of the heat. 'It was unbearable to go near them. The men who worked on the furnaces used to be allowed to drink beer.'

Pyrex goods were in demand and there were certainly perks to working for the company, Brenda explains:

> Now and again there was a sale on, and so you'd get Pyrex stuff for coppers. When you got married, you got a present and mine was a dinner service. I've still got a few of the plates after fifty-odd years. You could buy whatever you wanted, so my mam, mother-in-law and sister-in-law all got stacks of things when the sales were on. There were other factories around us too, and someone would come and

Brenda Forster (number 43) in the Glass Queen competition, 1961. *Brenda and Ed Forster*

> tell us there was a sale on at say, Mantle and Gown, who sold clothes, and we'd go down and buy loads.

Wear Glassworks and Pipeline tried to invest in their workers by putting on various events through the year. 'They used to have a sports club and have competitions,' remembers Brenda. 'We played old man's bowls. I remember the first time I did it, I rolled the ball and hit the jack, and celebrated, and I was told off because you're supposed to be quiet on a bowling green.' In a time when Miss World was still popular and beauty contests were the done thing, The Glass Queen competition came to Sunderland. 'Every year they did a Glass Queen competition. When I was 17 or 18, they told me to enter, and Miss White took some of the girls into the personnel department and got someone to teach us how to walk by putting books on our head! We were ever so …' Brenda does a little nose in the air, hoity-toity shuffle.

> The competition was at the rink in Park Lane, and on the night, I got ready and went down. Some of the directors from the factory were the judges. They had another little factory at North Shields, and the girl who won the year I was in it, was from there. She had lovely dark hair and loads of make-up on. My Dad used to say, 'you'll spoil your face putting make-up on, look at your mam's skin and she's never wasted money on make-up,' but this girl from North Shields, she had lovely eyelashes and hair, and a really sexy type of dress. She won, and then I was second and someone else got third place, so the two of us were the winner's maids. They took the three of us around the factory to show all the workers and then Miss White took us to the Marsden Grotto [a pub]; it was the first time I'd ever been and it was the first time I'd ever had scampi!

Outside of work, the entertainment in Sunderland was more routine and usually bustling with young people out to have a good time:

> We used to go to the rink in Park Lane on a Monday night and on a Saturday night we'd go to the Seaburn Hotel on the seafront and late night buses took us home. The hotel would be crowded and you'd be trying to jive in these little spaces; that's where I met Ed. We went to the pictures, but there were only about three or four picture houses in Sunderland. I never used to go to the ones they called loppy ones, the dirty ones. They were over in Southwick, so I never went to those in case I got fleas! There was a social club too, but it was a trek to get to Hylton in those days as there was only the Jolly Bus that would take you.

A 'Jolly Bus' at the top of Holmeside, 1965. *Sunderland Antiquarian Society*

I'd never heard of the Jolly Bus before, despite it only ending in 1995 – well within my lifetime – and have to ask Brenda what it was. She thinks it's the name of the owner of the fleet of local buses, but isn't sure and everyone just knew it as the Jolly Bus. I think this might be one of those names that you can drop into conversation with people of a certain generation and get an immediate response. I decide to test it with my aunt and uncle later, whose eyes brighten as they exclaimed in delight that yes, they do indeed remember the Jolly Bus, and go on to describe what they looked like (cream and brown) and how the back seat was a wooden bench, and if you sat at the back your teeth would rattle.

Typing it into Facebook, I discover that not only are people mentioning it often on local history groups that I'm a member of (apparently you could privately hire a Jolly Bus and jaunt off to Blackpool or somewhere equally glamorous), but there's also an entire group dedicated to it. 'Bring Back the Jolly Bus', boasts a less-than-impressive twenty members, but what they lack in numbers they make up for in enthusiasm, sharing pictures of the buses, old tickets and even anecdotes, such as this one from Scott Barrett: 'I always remembered the polished wooden seat at the back as a child. It was great fun to go round a corner on it as you would slide all over the place ... happy days.' Further investigation reveals that an original Jolly Bus

National Glass Centre, 2017. *Marie Gardiner*

still exists, and having spent some of its retirement at the North East Land, Sea and Air Museum (NELSAM), it is currently being restored to its former glory.

From our chat, Brenda seems to have thoroughly enjoyed her time at the glassworks; her eyes have lit up while talking about it, in a way that they haven't when she mentions jobs she's had since.

> I had a great time working for Pyrex, from being 15 to 28; it was lovely. I made nice friends and keep in touch them and still meet up regularly with some every month. Edna, one of the girls in the general office fell in love with Alan who was on the lathes, blowing instruments. They got married when she was 21 and I've kept in touch with her even though she lives in Canada now. When I was 15, I thought the men I worked with were old, and really they were only in their 20s, and I see them now sometimes and I still think they're really old!

In the strange way that these heavy industries seem to, things came full circle when American company Corning bought the remaining 60 per cent share of the glassworks in 1973. In the mid-1990s they (known collectively as Sunderland Glass Works) were split when Corning sold the domestic glass side of the industry, to Newell and Rubbermaid, and continued to manufacture scientific glassware. Newell sold to French company Arc International in 2005, who, later, after Corning announced its closure, made the decision to move production away from the UK to France as it was no longer financially viable, the factories having shared resources.

With the closure of Corning and Arc, glass production ceased on Wearside, ending a tradition that saw its roots in the seventh century. Considering glass production lasted the longest of Sunderland's traditional industries, it's surprising how little it's mentioned in relation to the shipyards or mines. In fact, glassworks are said to have employed the biggest workforce in Tyne and Wear after the shipyards.

At one time, everyone who had a Pyrex dish in their home also had a little piece of Sunderland's history. Today, our only reminder of Sunderland's love affair with glass stands in the form of its National Glass Centre, a heritage building commemorating a bygone industry, built in 1998 on the site of another – J.L. Thompson and Sons shipyard. It's an apt site, sitting close to St Peter's Church, part of the Monkwearmouth Priory where Biscop first introduced glass-making to Sunderland.

In 2010, the University of Sunderland took over the management of the centre and has worked to develop the studios and galleries. For the last decade, it's been free to enter and as such, has seen a rise in visitors keen to learn more about Sunderland's glass heritage. The university offers a degree programme in Glass and Ceramics, encouraging students at home and from all over the world to learn the skills of glass-making, although this is centred more on the contemporary arts than the practical manufacturing of old.

The website for the National Glass Centre states that 'up to 40 glassmakers' are working in the centre at any one time, and although a far cry from the hundreds of workers that used to turn out glassware for the country and the continent, it does mean that in some small way, we can hold on to our centuries-old institution.

5

SAFC

I've never truly understood the love of football, which is tantamount to a cardinal sin for someone born and raised in Sunderland. Football still played a large part in my childhood, though: friends at school were big fans, the girls as well as the boys, and part of your Mackem uniform required owning a recent Sunderland football shirt.

In my teens, I'd never been to a football match but I owned at least two Sunderland shirts. A group of us would often jump on the bus from Grindon and head to Whitburn's Charlie Hurley Centre, the training ground of the SAFC players. As the players left training and came out in their cars, they'd stop to sign our autograph books and pose for pictures. As I studied their scrawl on the page, I'd ask my friend Christopher who that particular player was, so that I could jot it down next to the signature. Kevin Ball, Michael Gray, Niall Quinn … these were the footballers from my childhood and I remember them clearly, despite never having seen a match.

The Charlie Hurley Centre has sat vacant for over fifteen years, since training moved to the Academy of Light. The gates to the centre, a firm and distinct memory of my childhood, were recently repaired and moved to the Stadium of Light, where they'll stay as a permanent fixture.

Despite having no particular love or affection for the game of football, there's something reassuring about those gates finding their new home.

Football seems to link us through the generations. While my contemporaries might not be able to comprehend spending twelve hours down a mine, or the deafening noise of the shipyards, we can all understand the concept of coming together to enjoy a game. It's hard to talk about the people working these jobs without also looking at how they spent their free time – or what little of it

Charlie Hurley Gates, Stadium of Light, 2017. *Marie Gardiner*

they seemed to have. Everyone I spoke to during the course of my interviews mentioned football. It was a community event, everyone was at the match and as Sunderland evolved as a town, so did its team, its stadium and the very nature of football.

As football is such a huge part of Sunderland's culture and everyday life, it seemed appropriate to start from the ground up and see if and how perspectives change. William Ernest Stout, or Ernie, has been a fan of Sunderland AFC all his life. Malcolm Bramley got his dream job with the club in 1962 when, as an avid fan of SAFC, he started working as an office junior. And Sir Robert 'Bob' Murray was on the board at Sunderland AFC from 1984; in 1986 he became chairman, a position he held until 2006.

Ernie and I talk at the Stadium of Light, where he still comes every other week during football season to support his team. I suspect his heart still lies with the old stadium, Roker Park, as his eyes light up when he remembers going to matches there. Like many fans, this great love of football, and in particular SAFC, seems to be passed down through the generations. 'I remember when I was 7 in 1937, my dad threw me up in the air, saying, "We've won the cup son, we've won the cup!"'

Like me and countless others, Ernie remembers the footballers of his youth. He rattles off a few names:

> Barney Ramsden, Tommy Wright, Tom Walsh. Ken Oliver, they used to call him Swan Neck Oliver, because he had a long neck and when he hit the ball it would carry on swaying. Arthur Wright was my favourite wing half, he came from Castletown and I used to love watching him do sliding tackles, and his wing to wing passing was brilliant.

I wanted to talk to Sir Bob Murray, to better understand the role that the club plays in the lives of the people of Sunderland. Getting a face-to-face meeting with him is tricky; he is, after all, an industrious businessman and still very prominent in the development of Sunderland through his links with the club. After some discussion with his assistant, Sir Bob kindly invites me to visit his home in North Yorkshire.

It's a stunning house, with sweeping views of the countryside awarded as an area of outstanding natural beauty. He jokes that the view would be better without the current, heavy mist blocking our outlook and insists I have a cup of coffee. I recognise a safety net of papers on the table in front of him, in case he can't recall something he needs to talk about. I do the same thing, placing my book full of notes in front of me, hoping that I don't have to refer to it. It's endearing to see that Sir Bob is not over-confident, and as he talks, he is softly spoken and considers each word carefully.

It's immediately clear that Sir Bob's working-class background is important to him. He is the only child of a fourth-generation Sunderland mining family. 'Nobody went to university,' he says. 'We all knew what we were going to be, you'd go in the mine – there wouldn't be another alternative. That area was very restrictive, you'd most likely go under, that was life and it was a hard life, but it was community driven and people looked after each other.' Sir Bob's father moved to Consett to escape the mine and to try and secure a different future for his family.

Although Sir Bob was born and raised in Consett, many of his family still lived in Sunderland and that, along with football, would keep pulling him back through the years:

> The first game I saw was in the mid-1950s and my father put me on a little stool in the Clock Stand so I could see over. We were playing Wolves that day and they were the best team in England at the time, they won the Championship. My dad told me that we'd drawn but when I looked back, we got beat!

I'm told football was very different in the 1950s and 1960s to what we've come to expect now. Ernie remembers the exhilaration of the game. 'We didn't win anything,' he laughs, 'but it was great football; they used to pepper the goal for about twenty minutes and it was all excitement, that's what I like.' Like Sir Bob's father, and as with so many generations, Ernie's love of football was passed down to his children as soon as they were old enough to go to Roker Park. 'I used to take about four or five of the family down and I used to make cradles for them – I was a joiner – I'd hook them on the wall, because they couldn't see. There was always a crowd of friends and family.'

In 1962, while Ernie was enjoying matches at Roker Park, Malcolm was starting his ideal job at the club. 'I left school at 15 and didn't know what I wanted to do,' he remembers:

> So I registered with the Youth Employment Bureau and one day I got a letter inviting me for an interview at Sunderland Football Club. By then, I watched all the football matches, all the players were my heroes, so this was an absolute dream. I went for the interview and there were two people in the office, George Crow, the secretary at the time, and Eddie Marshall, the assistant secretary. I walked in, and they both chain smoked so the room was full of smoke. George Crow put a piece of paper on the desk and said, 'Right son, in your best handwriting, write out the names of the players who played on Saturday, the score and who we're playing next week.' He asked me two more questions and then I was sent on my way thinking I hadn't done very well. A week later, a letter came inviting me to be appointed as office junior at £2 2*s* a week with an extra 50 shillings for working on a Saturday.

Seeing behind the scenes of your football club is something that gets no further than a tour of the grounds for most fans. To work there and at the same time have a passion for the team must at times be a conflict. Both Malcolm and Sir Bob agree that it's a privilege.

Malcolm still sounds overawed as he explains his first time on the inside:

> I'd never been behind closed doors, the inner sanctum. It probably took me a week or two to realise I wasn't dreaming. But you soon got used to it, and the reality was that the people you saw in those days as heroes were just normal people, friendly people. Many of them, because of my future career in football, I got to know as friends. Some have been friends for the last fifty years.

As an office junior and later, assistant secretary, Malcolm's job was diverse and he came to spend a great deal of time with the players:

Roker Park at the end of the 1963/64 season, going up to the first division; Malcolm Bramley, fourth from left. *Malcolm Bramley*

> We used to get thousands of requests for autographs, signed footballs, or signed shirts. I'd spend a lot of time going backwards and forwards to the dressing rooms, taking messages. Eventually I was involved with tickets, being responsible for all the complimentary tickets and I'd have players coming to me with all sorts of excuses as to why they should have extra tickets. I saw a lot of the players on a daily basis; two of my favourites were Charlie Hurley and Brian Clough.

Malcolm would go on to work with Brian Clough again, during a stint at Derby County, and the pair would become good friends.

Working alongside people in an industry with such big personalities was bound to yield some great anecdotes, and Malcolm doesn't disappoint:

> Brian Clough injured his cruciate ligament in 1963, and in those days, that virtually finished players. Somebody rushed up to the office and said we need an ambulance, so I called for one. When it came, I went downstairs and Brian

> was being carried through on a stretcher. He looked up and said, 'That's it, that's me finished.' He tried to come back to football and during the summer he would run up and down the terraces at the Roker end. I was in the tunnel one day and he shouted over to get him a ball and said get yourself in goal, I need to practise. So there I am, 17 years old, deserted Roker Park and Brian Clough the great goal scorer is firing shots at me from 30 yards like Exocet missiles. All of a sudden, a voice shouted, 'Bramley, get yourself over here!' I looked over and George Crow standing in the tunnel and he says, 'I don't pay you to play football, get yourself back to the office.' I said to myself what am I going to do, I've got the chance of a lifetime here. So I left it about five minutes and thought I better go back in. When I got back to the office, I got the biggest telling off of my life, but in fairness to the old boy, about two months later, he said, 'Do you remember when I called you in when Cloughy was firing shots at you? I would've done exactly the same, I'd have stayed where I was.'

Sir Bob Murray hadn't managed to land his dream job after school. He was, in fact, struggling to find work at all:

> I was born in 1946 and there were twice as many children born that year. After school, I was unemployed for a year because the steelworks could only take half the children in that year and give them whatever job, an apprenticeship, labouring job, a clerical job ... so I was unemployed for a year and that shook me up enormously.

Sir Bob threw himself into education, enrolling in Consett Technical College, and was able to get a job as an office boy at the steelworks after a year. 'I was studying as well as working, so I would go to the college and get an ONC in business studies, then I did A-Levels, went to Newcastle Poly on day release and finally studied accountancy.' Sir Bob climbed the career ladder, and when he returned to Sunderland in the mid-1980s noticed a distinct lack of prominent local investors. 'There were hardly any eminent families left, we had the Nicholsons and Sir Tom Cowie, Vardy, but there weren't many of them, they'd all lost their wealth or moved away. I've always found Sunderland to be a bit short of city fathers, which is a sad thing.'

At the time of Sir Bob's return, Sunderland AFC was struggling. 'I'd read there'd been some boardroom trouble at Sunderland, so I wrote to Tom Cowie and introduced myself, asked if there was anything I could do to help the club. Sir Tom asked me to come and see him, invited me to a few games and I joined the board in 1984.'

It was a bumpy start to Sir Bob's professional relationship with SAFC:

> Sir Tom took on a guy called Lawrie McMenemy, who was a disaster. Two things never should've left Southampton: the Titanic and Lawrie McMenemy! The club went from bad to worse and Sir Tom gave up really, so I took the reins. It was a difficult period, I was asked to sign the club into receivership at the same time Middlesbrough went in, so it was a really bad baptism having a potential liquidation to face and a manager called Lawrie McMenemy.

The one thing a true Sunderland fan looks forward to more than anything else in the season, is a derby with Newcastle United. Historically, Sunderland and Newcastle love to disagree. In the English Civil War, Newcastle was Royalist and Sunderland Parliamentarian. In the Jacobite Rebellions, Newcastle backed the Hanoverians while Sunderland supported the Stuarts. In terms of football, the first meeting of the teams was in 1883, so the rivalry goes back a long way. As such, sometimes tensions run high and measures have been put into place over the years to stop the two sets of fans from meeting, when passions are fired up.

In the days when the crowd was dominated by shipyard workers, miners and other industrial workers blowing off steam, there was no segregation of the two sides, but accounts of matches differ wildly. Malcolm remembers a more peaceful game in comparison to now:

> They were unbelievable. I'm not sure there was the underlying hatred that there seems to be now from a certain section, and not just younger people. I don't go anywhere near Newcastle when we play over there. You see so many different ages and it looks as though some people are just looking for trouble, I don't think that was the case so much back in the day; there must have been bits of trouble but nothing compared to what you can get now. I can't remember there being any significant disturbances, much less than there is now.

Ernie enjoyed derbies but tried to steer clear of any fighting that did occur. 'Opposition crowds used to mingle then and there'd be one or two skirmishes inside and coming out,' he remembers, 'but I used to keep clear of that, I wasn't a fighter.' An account of a 1978 Wear–Tyne derby from a Newcastle fan, on the site True-Faith, tells a much grimmer story: 'As a 14-year-old, I was in awe, as cars were trampled on, windows with Sunderland colours on houses and cars were smashed, with no resistance.'

Although the fans still love to taunt each other, violence is much rarer today, and with an increased presence from the police, we seldom see the levels of conflict and destruction that we have in the past. Some of these changes across the years can be attributed to the fans themselves. Where typically it would be the industrial workers flocking to Roker Park in their thousands, today, it's a different sort of pilgrimage to the Stadium of Light. 'We took the club from being very male to having the most lady season ticket holders in Britain,' Sir Bob explains. 'By taking a male club and slowly shifting it to a family club, we changed the whole culture.' Ernie has noticed the changes too: 'It's different now, there's no industry but they seem to come from all over. Back in the day they were all local people. I like the different cultures we have now.'

A key point with all three of my interviewees was the move to a more professional level in football and with it, the dramatic increase in wages. Part of Malcolm's role involved sorting out the players' earnings, and it's hard to imagine now, the picture he paints of professional football in the 1960s:

> I had to work out the players' wages, all on different contracts, different bonuses. It's amazing to think nowadays with the salaries that players get that in those days players were paid in cash. On a Thursday we'd ring for a taxi and I'd go to the old bank on the corner of John Street, collect all the cash, take it back and put it in little brown envelopes with the payslips. After training on a Thursday, about forty to forty-five professionals, would come along the dressing-room corridor and queue up the stairs. There was a little glass window and we had an old box that we put all the envelopes in, in alphabetical order, and they'd knock on the window, get their envelope and count it. Just think, if players were paid in cash today you'd need a lorry to carry it! It was only in the '70s and '80s that salaries started to rise and even then nowhere near the amount of money that players get now. But in comparison to the average working person,

Ticket sales at Roker Park. *Malcolm Bramley*

> for example if you took a miner's wage compared to a footballer's wage, they were very well paid.

As wages weren't the vast amounts we've come to expect today, the footballer and fan relationship was very different. Malcolm and Ernie both remember players riding to Roker Park on the bus with fans, they'd stop and chat in the street and the majority of them were local and lived in the area. Unlike players now where wages can set them up for life, many of the players had their own businesses, and those who didn't struggled to find something to fall back on when their careers ended.

Players today are traded like cigarette cards and few hold any loyalty to their team or home town. 'The lads who played for Sunderland, the likes of Charlie Hurley and Kevin Ball, they're Sunderland diehards, not like Darren Bent,' Ernie scoffs. 'We were playing Newcastle, the biggest game of the season and he had his resignation in his back pocket when he came. I've got no time for those people.'

As favourite players of bygone eras go, Charlie Hurley is a name that crops up time and again. He played for Sunderland from 1957 to 1969 and was named Sunderland's 'Player of the Century' by fans at the centenary of the club in 1979. The training centre was named after him, and he's a firm favourite with Sunderland fans. Malcolm had the privilege of knowing him personally during his time at SAFC and recalls an amusing story:

> Charlie always asked for extra tickets, so I used to help him out. I used to babysit for his kids too and he'd take me home when he got back in his brand-new Ford Zephyr, and I loved that car, it was the kind with bench seats and the gear stick on the steering wheel. One day he said, 'You've helped me out with tickets so I'll do something for you, I know you like my car, would you like to borrow it for a weekend when I've got an away match?' A few weeks later he came up to the office on a Friday afternoon just as they were about to get on the bus to travel to an away match, and said 'Here you are, bring it back on Sunday night.' I'd been asking a local girl out for ages and she didn't want to know, so I rang her up and asked if she'd come out with me that night, and she said she wasn't interested. Well she was a massive football fan, so I said I had Charlie Hurley's car … there was a huge pause and she said 'Pick me up at half past six.' When we got back, I dropped her off and was reversing the car when suddenly there's an almighty crash and the neighbour's lower wall disintegrated completely. She ran into the house and I panicked and drove home. She rang the next day and said the neighbours have got the registration number of the car and if I didn't come back and sort it, they'd ring the police. That was bad enough, but then I had to go and knock on Charlie's door.

> He asked how I'd gotten on and was everything okay. I said 'not really ...' and we went around the back of the car and there was hardly anything left of it. He went crazy, swearing at me left, right and centre and said I could babysit his daughters until they're 35 but I'd never get anything else from him. He slammed the door and I thought that's my relationship with Charlie over.
>
> On Monday morning I go back to the office and one of the players says, 'Charlie wants to see you in the dressing room.' I went in and there were about forty players sat around. They looked at me and then all of a sudden they started singing, 'Baby You Can Drive My Car' by the Beatles. Charlie had put them up to it. He took it in good spirits, but he never loaned me his car again and the girl never went out with me again.

Even as a self-confessed 'non-fan', I've still felt the contagious excitement that comes from watching an important match in a room full of fans; likewise the disappointment. Slipping out of the top tier is something dreaded through the generations. Ernie remembers the lows of 1987, when Sunderland were relegated to the third division for the first time. 'I was just about crying,' he remembers:

> It takes that long to get back. I think Sunderland deserves a Premiership team, we must stay in the Premiership; I'm too old to have to wait a few seasons to come back up again. It's important to Sunderland people, you're left with the sob story of thinking about next season, you dread going down.

I was curious as to whether working at the club would affect the way you enjoyed the game. According to Malcolm, it made you even more nervous:

> Everything was about winning games and the impact if you lost, or if the club were lower down in the league then there was potential for relegation and people losing their jobs, so it almost had more of an impact than when you were just a normal fan because it affected you from day to day. You saw the attitude of people on a Monday for example, after we'd lost a game, the atmosphere was completely different. The club and football in general was everything. People lived and breathed it and it affected people when they went to work, whether they were in the shipyards or the mines or whatever and they still do now.

I think at some point most people have wondered why English and, in particular, northern football is so passionate. Why, even when the team is on a losing streak, when they play badly or seem to let their fans down, do people continue to support them? Malcolm believes that even in the days of heavy industry, support wasn't just through lack of choice:

> There weren't the alternative attractions, but then conversely you could say nowadays there are so many attractions but the club still attracts tens of thousands every fortnight with little experience of any success over many, many years. I think partly it's a cultural thing, for example I'm a parent and I brought both my sons up to be Sunderland supporters, they had no choice. One lives in Australia, the other lives locally, but they're both absolutely fanatical about Sunderland.

Part of the great love of football can be seen to stem from a lack of opportunity. This is highlighted around the world in areas of dire poverty like Brazil and Africa, where the playing of football can be the quickest means of escape and from where new talent emerges regularly. Watching football can be aspirational for many; seeing someone who has vastly improved their quality of life gives hope to those in a town whose economy is struggling and where unemployment is rife. Football is also an escape, and I think back to Albert the miner who absolutely hated his job, and how fondly he spoke of going to the football matches – it was a distraction from the thought of your limited time in the sunlight before another twelve-hour shift of darkness loomed, or replacing the relentless clanging of the shipyards with the roar of an excited crowd.

Sir Bob understands how much football can mean to someone who has struggled in a poor town:

> It means so much, it's such a big part of their lives and they care passionately about it. It's not like some of these London clubs where they turn up when they want. Sunderland and the fans were looked on in great esteem. Everybody respects Sunderland fans, they're knowledgeable and they're passionate about their team. This isn't an ordinary club, it's a special club because you know how much it means to them. I'm one of them, I was unemployed for a year, I've lived on low incomes, I've had life without a car and not been able to do things because I don't have money. It's a big responsibility being chairman of that club. It's the biggest thing in the city, it's a proper football club. These are the best people I've seen in football. I know from time to time I've been unpopular, but if you're in a job for twenty years and it's as passionate as this, you are, it goes with the T-shirt. They're first-class to me and it's a great thing.

Sir Bob mentioned Sunderland being different from London clubs, because they turn up come rain or shine to support their team, and it's true. Many friends of mine

An aerial view of Roker Park, 1967. *Tyne and Wear Archives*

are fanatical Sunderland supporters, and even when things aren't going their way and you can see they're furious, they still band together and support them. Eddie feels the same way:

> I love going but I can be critical with them as well, because we've seen some bad times and you think 'Why am I going?' People of Sunderland deserve a Premiership club, it's part of us, you've got to have a good team round here, Newcastle is the same. They say 'Why do you go?' and I say you can't let them down now, it's like letting an old friend down.

The love of football is a hard thing to explain or quantify. When I ask everyone why they support Sunderland or why they like football, they just can't articulate a reason. 'I suppose it's in you,' Ernie laughs, and this isn't the first time I've heard that. 'It's in your blood' is a common turn of phrase for explaining the sometimes cult-like support of a football team. Ernie remembers having to decide between going to the match for a time and being able to watch it from the comfort of his own home:

> I love the cup. I've been to Wembley, but I didn't go in 1973 because my son wanted a colour television. I went and put a deposit down, and I said it must be here on Saturday morning. A friend turned up on the Friday night, with a ticket to the match and I couldn't afford it because I'd paid the money for the television!

When Sir Bob Murray's father took him to his first match at Roker Park at the age of 9 I doubt either of them imagined that somewhere in his future, he'd knock the beloved football ground down:

> Roker Park was a great servant. The first fifty years were good, they had a club to envy, they could compete with the best. Before money came into it, they could rub shoulders with Arsenal, they could win the Championship, they could play at Wembley, they got great players and they could keep them. Those times went when the professional side took over and the maximum wage went. All the people who had hard working conditions and short lives, could go to this place and have the best on their doorstep.

As football evolved and ground requirements and laws changed, it was becoming clear that Roker Park was no longer a suitable home for Sunderland Football Club. Attendances had plummeted. Sir Bob remembers attending matches with huge crowds that had now dwindled:

> When I took over the club we had a crowd of about 8,000. The year before we came up, the last season at Roker Park, our capacity was 23,000 and we only sold out twice: once against the team up the road, and once against Manchester United. This big club couldn't sell out in what was the Premiership, Division One; that time had gone.

After the tragedy of the Hillsborough disaster, where 96 people were killed and over 700 injured, the Taylor report recommended that football grounds convert to all-seater, signalling the beginning of the end of Roker Park:

> The last fifty years of its life weren't so clever and it needed a new beginning. The Stadium of Light has given it that and it's a Premiership stadium in every sense. It's been a steep change. We haven't got the biggest stadium, but that stadium can easily be turned into a 66,000-seater. The infrastructure, the land, the development

The Stadium of Light and Pit Wheel to commemorate Wearmouth Colliery, 2017. *Marie Gardiner*

> pattern, it's all there if somebody came and was really successful. When things are going well here, the noise in that stadium is scary. I was at Wembley when we played Manchester City the other week and I know the people at Wembley; I used to be on the board, well when our goal went in, they said that was the loudest roar they'd heard since new Wembley was opened.

It wasn't just the management who agreed it was time for a new stadium; the fans were behind it too. Malcolm thinks it was a welcome addition to the city. 'As fond as many of the fans were of Roker Park, most agree that the time had come for a new stadium, not least because it was one of the few to move closer to the city centre, making it more accessible.' Malcolm thinks the atmosphere has changed though, and not always for the better:

> A lot of the new stadiums, partly because of seating, changed the atmosphere. There's still a good atmosphere at the Stadium of Light, but there was a closeness at Roker Park, like most of the old grounds; the pitch was nearer the spectators, so there is a difference, but on some nights here I've seen atmospheres that do come pretty close. I think back to the days at Roker Park and there are so many differences, like the pitches … pitches now are made of different sorts of grass, players are basically playing on bowling greens compared to the mud heaps that they used to have; the catering facilities, the comfort – a lot of old grounds, you had pillars, so people couldn't see. From that point of view it's progress and good progress.

The Stadium of Light has given a new lease of life to Sunderland, in some ways becoming its own industry; bringing money and business to the city and boosting its reputation. As well as a football stadium, it also has a banqueting suite and several conference rooms that can be hired for events. In 2009, it became a concert venue for major artists and has hosted Beyoncé, Take That, Coldplay, Rihanna and more. The stadium is the home of the University of Sunderland's graduation ceremonies, earning it the award of 'Most Creative Use of a Sporting Venue' by *RSVP* magazine.

The site of the Stadium of Light, Stadium Park, is now also home to a number of new businesses, each with their own unique opportunities for Sunderland. Sir Bob is very passionate about the recent developments: 'We brought the stadium, then the Aquatic Centre, The Hilton Hotel and now we're bringing the Beacon of Light, and I hope there's more to come because it's a destination site.'

Sir Bob shows me his plans for the Beacon of Light, an education, training and sports facility, and you can see how immensely proud of the project he is, and of Sunderland itself, 'I love the city and I love the people. The things I've done for the city haven't been for personal gain, but because my heart's here.' With the ups and downs of any football club, it's easy for a former chairman to be less than popular with the people, but it seems like Sir Bob has remained on good terms with Sunderland, having even been awarded the Freedom of the City:

> You can't go to university to learn about football, you've got to buy the T-shirt and wear it; it's really, *really* difficult. I was on the board for two years, then I was chairman for twenty years and I would do it all again, I'd do it all again. We had great times and a lot of success, it was us all coming together and I've only got fond memories.

6

The Docks and Port

Sunderland's docks and port are at the heart of the city's industry in many ways, but I also have a personal connection. My granddad worked at the docks for thirty years, following in the footsteps of his own father, as was the tradition in many families.

I get in touch with Jack Curtis while looking for a docker to interview. He explains that as well as being a local historian, with specific interest in the docks, he has a unique perspective on them as he also grew up on the docks in the East End. I go to his immaculately kept house and am greeted by a tall, straight-backed man with a booming voice and friendly face. He shows me a picture of his grandfather on the wall in his study, who is the double of Jack. 'My old grandfather looked after all the underwater maintenance on the docks,' he tells me:

> The sluices, the hydraulic systems to move water from one side of a sea loch to another. He was the superintendent diver for the underwater foundations of the South Pier. When he started diving, the river was partially developed and the docks were developed, so most of his work was maintenance and routine repairs, as we suffered a lot of storm damage.

Sunderland grew up around its river. As trade, and in particular coal, advanced, the river facilities were no longer fit for purpose and so merchants appealed to parliament for improvements. Unfortunately, traders on the Tyne feared that developments to Sunderland might impact negatively on their own businesses, and so the request was denied.

Jack explains the introduction of the River Wear Commissioners: 'It was formed 300 years ago, in 1717. The foundation of the docks was really started by the commissioner-appointed engineers, because it took civil engineering to develop the river.'

The Newcastle merchants once again opposed the bill to make improvements, but this time without success, and plans were put into place to create a new channel up the river. One of the first things the commissioners implemented was the original South Pier, which was completed in 1723. By the mid-1780s, they were working on another pier, the North Pier, which took a whopping ten years to build, finally finishing in 1796.

Sunderland got its first proper lighthouse in 1803, replacing the reflector light. Despite the major changes occurring, still more improvements were required. Jack explains: 'The river couldn't handle the amount of coal that was produced in the Durham coalfield. The necessity arose for either more facilities in the river, or the building of a dock; or the trade would go to the Tyne or the Tees.' In 1828, work began on the North Dock, designed by Isambard Kingdom Brunel. 'The problem with that,' Jack says:

> was the coal was on the south side of the river, and there was no bridge across. Brunel actually designed a bridge to carry coal only, but the whole scheme fell though. What we wanted was a dock on the south side, but to excavate into the land, which was high land above the river level, was going to be a terribly difficult job.

By the time the new dock opened in 1837, it had taken so long that it was already too small to meet requirements and notoriously difficult to navigate.

Sunderland Dock Company was formed in 1846 and John Murray, the port engineer, started to look into developing a South Dock in consultation with civil engineer Robert Stephenson:

> They were having problems with the big cliff that ran along the coast, it was about 35 feet high and suffering from a lot of erosion. Murray decided to put groynes [low walls or barriers] out into the sea at right angles to the land. They allowed the dross [rubbish or debris] that was carried by the tide to fill up between them.

They started to dig out the rock and tip the spoil loads on to the growing barrier, using temporary dams to hold back the water. 'The South Dock was unique in as much as most docks were excavated out of the land, this was won from the sea; a brilliant engineering feat,' exclaims Jack with passion.

In February 1848 the foundation stone for the half tide basin was laid, and the dock opened in 1850. 'There were between 600 and 900 men employed continuously in the building of the docks,' marvels Jack. 'Murray developed the docks specifically for the shipment of coal. Coal was needed to boil water, to create steam and steam created the industrial revolution, so the whole thing hinged on the ability to provide coal, to finance the industrial revolution.' The South Dock was extended in the mid-1850s and by 1860, Sunderland had increased its coal exports by more than half.

A new lighthouse was built for the South Pier in 1856, by Thomas Meik. It's still around today, but rather than adorning a pier it now sits proudly in Cliffe Park, where it has since 1983, enjoying its retirement with a sea view. Meik was a busy chap, and as well as building a new lighthouse he also constructed an entire new dock, which was named Hendon Dock. In 1868, Henry Hay Wake took over from Meik as chief engineer and by the late nineteenth century had started working on plans for a new harbour and piers. Plans were approved, and construction of the new North Pier, now known as Roker Pier, was started in 1885. The pier, freshly

The view from Lambton Staithes, 1972. *Sunderland Museum and Winter Gardens*

East End, Hendon, 1955. *Sunderland Museum and Winter Gardens*

adorned with our much-loved red and grey lighthouse, was opened in 1903.

Jack remembers a thriving port as he was growing up in the 1940s: 'The docks, when I was a boy, were heaving. The whole west side of the docks was occupied by coal staithes, known locally as "the drops".'

> The coal was brought there in 21-tonne coal trucks, put on to gradient lines and secured. They were then unhooked one at a time and run down the gradients on to the coal staithes, discharged into a big hopper and from there, down a big chute into the hold of the ship. There were various trades at work: teamers for teaming coal, trimmers who went into the hold of the ship and trimmed the coal so it was stable at sea; there was as shipyard and an engine works there, which was largest in the country when it opened in 1868, that was the North Eastern Marine Engineering Company, and my other grandfather, my paternal grandfather, went there as a 14-year-old boy to serve his time as a marine engineer, so I've had a long association with the docks.

Growing up on the docks must have been an incredible experience, but growing up there during the war was something else entirely. 'I carried my grandmother's groceries down every week during the Second World War, from the Co-op store in High Street, down onto the docks, for the entire war. I was the only schoolboy in Sunderland who had an official dock pass,' Jack remembers. Of all the vivid memories he still has of the docks at that time, one stands out above all others:

> In 1941 or '42, there were two frigates built in Canada and brought here to be equipped with guns, depth charges and so on. They moored them at the Boiler Shop quay at North Eastern Marine, in the dock, which was about quarter of a mile from our house. Eventually, they decided the ships were far enough along to bring the crew over for familiarisation. While they were here they were allowed to have a shore party to come into Sunderland to enjoy the delights of our town.

He chuckles as he continues:

> The first night they went ashore but when they came back, maybe a bit tipsy, one of them fell overboard into the dock and was drowned. A couple of days later, the same thing happened again and another one died. My grandfather came into the house and said to Grandmother, 'You'll never guess, another one of those lads drowned in the docks last night, I had to go with the grappling irons to find him; that's the second one.' My grandmother had a bit of a think and said, 'They've come here to fight for us. Get along and see those two skippers and tell them we'll meet the last bus along the top of the dock bank every night and we'll put the shore party back on the ship.' They did that until the ships sailed and no more sailors drowned. Every week, my grandmother got a box of groceries from the ships, which were like gold during the war, when we were all on rations. There was a tradition in the Royal Navy: when you were in service you got a ration of rum, but you could have 'sippers', which meant you saved a sip, and didn't drink it all. The two crews went on sippers and the rum they saved every week came to Grandfather, so we had a house full of rum by the time they sailed away!

Jack's links with the docks were obviously close, and with the benefit of hindsight it's easy to see how they've shaped his life and his interests. He's not only knowledgeable about the history of the docks, he's passionate about it: 'I had a university education while I was still in short trousers.' He laughs:

> I completely failed the 11-plus because I took that in 1941, in the middle of the war. I'd been evacuated and I didn't like it so we came back, but we lost an awful

> lot of education. I had a wonderful boyhood, because I had the run of the docks. I could stand at the engine works and watch the turner on the big lathes, turning a tail end shaft, the brilliant steel spirals coming off like curls. I could go into the shipyard and see all the different trades working. The noise was tremendous, there were tugboats chugging in and out of the river into the docks, there were ships moving on the tide, they were towed out through a half tide basin into the river itself. During the war in particular, they formed up in convoys off the port because of the danger of U-boats and such. The whole docks were a hive of industry: saw mills, chemical works, engine works, storage depots, cranes everywhere, little engines running about and beautiful swing bridges that opened and closed with the smoothness of silk.

I feel exhausted just listening to Jack talk about the comings and goings of the docks. He paints a vivid picture, but still one that's hard to visualise with today's context. 'I can remember the dockers carrying hooks; they had a circular hand with a metal hook through, and they used it when they were handling bails of sisal,' Jack says. Sisal is a fibre used in the making of things like rope, cloth and carpets:

> We used to import sisal from East Africa. It would come in big bails and they used the hooks to move it onto the trucks, then into storage. It was then taken from the docks up through Sunderland, over the bridge and down Roker Avenue to the British Ropes factory, where it was spun into ropes. The docker would work in exposed conditions, in all weathers, in conjunction with crane drivers, because it was cranes that had to lift the cargo out of the ship, into the trucks to take it wherever it was going.

The docks and other industries fed from one another. While they shipped coal away from Sunderland, they also brought in pit props to secure the mines. Jack explains:

> They stacked millions of pit props from Scandinavia in great 30-foot columns. The whole dock was covered in stacks of pit props and the mines called them off when they wanted them, then they were loaded into trucks again, put into the rail system and taken to Ryhope Colliery, Silksworth, Wearmouth; all the collieries in the Durham coalfield. There were brownish-red trucks with the name emblazoned on the sides: Lambton, Hetton and Joicey Collieries, that brought the coal away from the inland pits by rail on to the docks. They were filled up with pit props and they were taken back to the pits. The whole place was on the move. It was dangerous, though. I went to work at 14 and I was warned, you don't

A docker loading up a truck, 1966. *Sunderland Museum and Winter Gardens*

run, you watch what you're doing and listen to what you're being told. You had to learn to be careful.

The docks figured large in the prosperity of Sunderland, there were all sorts of trades employed on the docks, the whole thing fit together like a jigsaw. British rail was coupled into the dock, but in those days it was the London and North Eastern Railway. They had an incline feed into the dock and the River Wear Commissioners themselves had their own little tank engines which moved trucks all over the docks. It was rail that opened the docks in Sunderland, not road transport as the roads weren't very good in those days.

The development of the port and docks seemed to come on in leaps and bounds until the 1960s, when the decline of shipbuilding and the collieries slowed it dramatically. The busy, vibrant place that Jack remembered from his childhood became unrecognisable. 'The whole thing changed dramatically,' he says sadly.

'We lost an awful lot of trade and the docks became a quiet backwater in my opinion.'

In 1972, the Port of Sunderland Authority took over from the River Wear Commissioners and their meeting in September of that year was to be the last in their 255-year history. My grandfather would undoubtedly have seen much of this change, having worked at the docks from the mid-1950s, for thirty years. It's not something I can ask him about, as he died some years ago, so I set out to find a docker from the same era. Via a Facebook group, a man from New Zealand contacts me and tells me his father-in-law worked at the docks from the 1970s. I ring Alan Gregory to chat about an interview, explaining what it is I'm looking for. I'm completely stunned when tells me that he used to work with my grandfather and knew him well.

I arrive at a terraced house in Pallion and am greeted by a huge barking Alsatian being held back by Alan. 'He's soft as anything, don't worry,' he shouts over the dog. I manoeuvre gingerly around 'Troy' and make my way into the living room, where I meet Alan's wife, Edith. Perching on the edge of an armchair, I'm comforted to see that Troy has settled into a corner, although he's still watching me with interest.

I know that my granddad was a docker most likely because his father was; I was curious to know if it was the same with Alan:

> Aye, my father worked at the docks from leaving school, and his father was a foreman on the docks before that, and his father before *that* – as long as there have been dockers in Sunderland. When my dad was there, 200 men worked down the dock and they had to turn up with their work-books in their hands. If the stevedore wanted twenty men he'd pick that one, that one.

Alan gestures, jabbing his finger at an invisible crowd:

> Whoever he wanted, whoever's face fit. If it didn't fit, you weren't working and you got sent home. You had to be back down the dock for quarter to one for more of the same carry on. It all depended who the stevedore was and who his blue-eyed boy was. My dad didn't work for eighteen months because he had an argument with a stevedore; he turned up twice a day for eighteen months and he got no work.

'Stevedore' has its roots around Portugal and Spain and came to be used by us via sailors spelling *estivador* or *estibador* phonetically. Loosely translated, it means a man who loads ships and cargo. What a stevedore was to the dock varied from place

Low Quay, early 1930s. *Sunderland Museum and Winter Gardens*

to place, but in Sunderland it was the master of a loading gang; the boss. In a time before equal rights and fairness were considered, getting on the wrong side of the stevedore could find you out of work for a very long time.

My gran mentioned recently about a time when the docks were so short of work that they told the dockers to go and get other jobs and they'd keep their positions open for them when more ships came in. Rather than have my granda go to a factory, my gran got a job in Pyrex – she said working indoors would have killed him. This seems to have been a common feeling among men of the heavy, industrial trades; they needed to be outdoors. Alan says it was the same for him:

> I was an outdoor man. Who wants to sit in a stuffy office all day? When I was on the tugboats at Seaham Harbour, I got one day off in twenty-one and on my day off, I used to go fishing. Snow that deep on my head, but I'd still go. I'm an outdoor man. I was working at Seaham Harbour docks and travelling all the time. When the jobs came up in Sunderland, I put in for one and got it – it was all through family. You'll find that the same names reoccur, right back into the past. My first day was 7 December 1970 and I turned up at the control point. What you were loading and unloading would depend on which ships were in. It could be sisel, cement, corned beef ... anything!

In 1945, there had been a strike at the docks for an increase in basic pay. The response to this was the 1947 Dock Workers' Regulation of Employment Scheme, where a local board was given control of workers, wages and so on with the view of guaranteeing dockers a right to minimum work. Although work no longer relied on the good graces of a stevedore when Alan started work, the labour was still on a casual basis. 'If there was no work you'd come down to basic wage which was the equivalent of £5 a week, not even that, to be honest,' he says:

> You still had to turn out twice a day for that. You could be working one day and if there was nothing the next, you still had to turn up at eight a.m. The boss would pick the telephone up and ring the speaking clock; you could be fifteen seconds past eight but any later, you'd be sent home and lose your day's pay and your turn for work. When you were working, you got 'tonnage', paid by the tonne. We worked on a rota system: when you finished your job, you went to the back of the queue and the rota would move up until another job came in for you. We only had one condition, you couldn't be paid off unless you were caught stealing. You could tell the manager where to go in no uncertain terms, but if you were caught stealing you were sacked.

Dockers worked in gangs, which were often grouped together alphabetically. It makes sense that my granddad, a Gardiner, would have been put together with Albert, a Gregory. Each gang would usually be made up of four or five workers, including a supervisor, or 'boss'. Gang members have particular skills or jobs, such as lashers, swingers or swingmen, and gangs could have intimate knowledge of different kinds of ships. If a gang is missing a member – through sickness, for example – it can impact the amount of work they can do and the efficiency of that work. A complete gang, that works well together, is a valuable asset.

I ask Alan what it was like when they worked with my granddad, and Edith chips in from the sofa: 'You were called the Greedy Gs, weren't you, because you worked so hard?' I've not heard this before and it's intriguing – the Greedy Gs?

> We were the only complete gang on the dock and we'd do 370 tonnes of cattle cake. Cement boats were 500 tonnes a day and you handled every bag of cement: we took them off the pallets and put them in the ship's wings. You got paid gang for gang if you had a sisel boat, and you'd have four gangs on. If they did well, you got paid well, if they didn't then you got paid what they'd done. You flogged yourself to death to get the tonnage out. We had to make hay while the sun was shining.

Being a docker has always seemed like gruelling, tough work to me. Being paid in tonnage meant that dockers were constantly working as hard as they possibly could, to make as much money as they could, just in case rainy days were ahead. I wondered if it was something you just got used to, or whether you felt it every day of your working life. 'It was hard work, you had to be fit,' stresses Alan:

> You had to pass a medical and we went down to Hull for three weeks to learn health and safety, how to drive forklifts, do knots, lashes, slinging and all that. You had to be trained before you went down. Some of the jobs were horrible. If you were at the front of a ship, you'd carry wet animal skins over your shoulder, and you'd have maggots on you as big as your thumb. All you got was a bucket of water to wash yourself down with.

Edith pipes up: 'He's come home and I've had to bath him in the yard because he was covered with maggots.' Alan continues:

> If you worked on a coal boat you came home black, there were no showers or anything. When you worked in rice polishing, you'd get covered in insects. Little

> Billy Harper and me walked off the dock at seven o' clock one night and people were parting like waves and staring because the insects were all over us. I took my overalls off in the yard one day to lash all the insects off, and I was hitting it against the wall, and my wages came flying out of the pocket and over the wall. The bairns were running down the street trying to pick my wages up!

As with all workplaces filled with potential hazards, accidents were fairly common at the docks. Thankfully, my granddad was never seriously injured; the worst he suffered was a narrow miss, when a tailgate hit him and cut him just above his eye. Alan, however, wasn't so fortunate:

> I got crushed swinging the grab. We were in a kerphalite ship at Hendon and the crane cut out. I got squeezed against the tunnel and my pelvis was crushed; I was off work for eleven months. I went back as a light duty man but there weren't many light duty jobs, so I wasn't getting any work. I went as a fit man instead and on my first day back at work, I was lifting 16-stone bags. I came back with blood running down my legs. All I got was £500 for pain and suffering and I've suffered ever since. We didn't get sick pay either. I went back and worked on the ship I was injured on when it came in again and one of the lads said, 'I don't know how you can do that,' but you have to, I was a young man with two kids and a mortgage.

Aside from loving the outdoors, it seems like conditions on the dock were fairly awful. The dockers worked in all weathers, doing gruelling physical labour and trying not to get injured or killed. Aside from necessity, which I fully understand, I couldn't comprehend why Alan seems to have actually enjoyed his job. 'It was a good job,' he insists:

> You worked in gangs, so there was laughing and carrying on as you worked; it was a happy job. You had to have a sense of humour, though, no matter what you said there was always someone to pick it up and make a joke. When you're working a job like that it's all about patter. We had nights out together, we had a welfare, and we'd organise concerts, days at the races, all sorts. You'd organise a bus and get £5 or £10 spending money, depending how tight things were at the time. It was a great community. It never entered my mind to pack it in, until Margaret Thatcher cut the cable. I loved it, and everybody that was there really enjoyed the work. It was the camaraderie involved, you were always laughing and carrying on with each other.

This sense of community and looking out for one another at work seems alien today. 'When the men got older and couldn't manage, the younger men would carry them,' Alan tells me:

> They got the same pay as you, for all they couldn't manage, but you worked for each other. Little Billy Harper was getting on a bit – the iron ore boats were big – Billy got halfway up the Jacob's ladder [a ladder used to allow access over the ship] on the way out and he got stuck, couldn't get any further. I climbed from the bottom ladder, put my head through his legs and carried him up to the deck. We rolled over the top and we were lying on the deck ...

He mimes panting and laughing. 'That was life, you did that sort of thing.'

Alan turns serious:

> Not many lasted to 65 and within two years of them retiring they were dying off, not very many had really long retirements. The type of work they did knocked hell's bells out of them, and it was never appreciated. They'd give you a job on Christmas Eve and say if you didn't turn up you'd lose your Christmas pay, and if you didn't turn in after Boxing Day, you'd lose your Christmas Day and Boxing Day pay. It was never, 'Merry Christmas, have a good time,' it was a warning, turn up or lose your money, and that happened all the way along until retirement.

The introduction of shipping containers and the development of technology were largely to blame for the decline in the need for dockers – nationally, not just in Sunderland. Alan remembers the move from hold storage to containers:

> Ships would come in and we'd do 1,100 containers off and 1,100 on in two days, working round the clock. Mechanisation was the killer, you didn't need as many men so the workforce just went down. It went from 270 of us, to 60, then later, down to 30-odd. Liverpool went from 28,000 men down to 9,000, then to hundreds; big docks, like Cardiff, closed down.

Part of Sunderland's problem with the increased use of containers was that the docks just weren't able to handle ships of that size. Jack explains:

> When the dock was first built, the alignment of the gateways weren't in a straight line, they're on different angles. This restricts the size of ships that you can get into the dock. If you want big container ships, the only way you'll get them in is to

Corporation Quay then (1949) and now (2017). *Sunderland Museum and Winter Gardens and Marie Gardiner*

> modify the gateways. When the dock was built, the regular ships using it would be between 100 tonnes and up to 1,000 tonnes at the most, so in truth the design criteria was totally different to the requirements today.

As dock work became scarce, the men found themselves increasingly without work. Alan remembers how difficult it was:

> When the work started to decline, you were down the dock early morning to get your name called out – no work – you got sent home. You had to be back for one o'clock to get your name called out again – no work – and back home. You just accepted it was part of your life. That was your job and you accepted it, and you waited for the next ship to come in.

As with many major British trades, Margaret Thatcher's policies regarding nationalised industries hit hard at the docks. The National Dock Labour Scheme was scrapped in 1989, meaning dock workers no longer had any guaranteed income, not even basic pay. 'Margaret Thatcher finished us,' says Alan forcefully:

> All the dockers in the country were finished. We all got our golden handshake and went. The contract the council offered us wasn't worth the paper it was printed on. They wanted you to be available from a minute past midnight on a Monday morning, to a minute to midnight on the Sunday. If you started on a ship, they wanted you to work it until you finished. They were rubbish conditions, so none of us accepted it so we all finished on the same date. We were told on Friday afternoon, 'Go home, you're all finished on Monday.' We turned up on Monday, were marched on to the dock one at a time into the control room, given our papers and that was it. It happened all over the country. If a ship came in, they'd pick up men from the dole for casual labour – no training, no health and safety.

Today, the Port of Sunderland primarily deals with scrap, steel and construction aggregate. The docks that were so vibrant with activity and life are far more sedate. While you can often still see a ship berthed at Corporation Quay, or watch the comings and goings of the port from the roof of the National Glass Centre, the trades and industry remembered by Jack and Alan are long gone. Now, the riverbank hosts a mix of student accommodation, restaurants and offices and the Port of Sunderland is, in the words of its website, 'striving to meet the challenges of the 21st century and demonstrate that it is still a key asset to the city of Sunderland'. As it reaches its 300th anniversary this year, some might say it's not doing too badly.

7

Shipyards

If you ask anyone in Sunderland what we're famous for, nine times out of ten the reply will include shipbuilding. Starting with the building of fishing boats in the mid-1300s, the industry grew on both sides of the River Wear, and by the mid-seventeenth century shipbuilders were turning out colliers, vital for moving the vast amounts of coal the North East was turning out. The coal trade was expanded to meet the demand of iron founders producing guns for the Napoleonic Wars, and Sunderland really hit its stride, producing around ninety vessels in 1815.

The first iron ships in Sunderland were built in 1852, at which point there were some sixty-five shipyards. Shipbuilding dipped in the mid- to late-1880s with the first of what would be three great depressions. Unemployment was rife, and those who managed to keep their jobs saw a drastic reduction in wages. The second depression, due to a national fall in ship production, was in 1908–10, shortly before the First World War. It picked up again when the war started, with yards such as the one at North Sands building cargo vessels and Admiralty craft. It was a victim of its own success: the third depression occurred with a drop in the demand for ships after the war ended.

By 1920, as building costs increased, the yards were once again starting to struggle, and for the first time in its history, in 1922, North Sands failed to launch a ship. The following year, five yards had failed to make a launch and the boiler-makers' strike of the same year was delaying production at others. Yards faded out of existence and left a total number of seven on the Wear by 1935.

Under the government's British Shipping (Assistance) Act of 1935, which subsidised the shipping industry, the UK Scrap and Build Scheme was brought into effect. This scheme was designed to reduce the British merchant fleet while

The launch of *Empire Bruce*, at Sir J. Laing & Sons, 11 June 1941. *Sunderland Museum and Winter Gardens*

providing some much-needed work for those companies who could scrap two tonnes for every one built.

The shipbuilding success that most speak of now is usually in reference to the Second World War, when Sunderland was making a significant contribution to the war effort, but things were, comparatively, still far from those of its heyday in the nineteenth century. Technology had improved, more ships were being built, but there were fewer yards and less people being employed. However, things had picked up, and with eight new yards added to the once-again thriving riverside, Sunderland started to produce lifeboats, whalers and navy vessels. The end of the Second World War paralleled its predecessor and strong overseas competition saw many of the major firms begin to merge in the mid-1950s.

William Hilton has been known to friends and family as 'Paddy' all his life, the nickname having been passed down through the generations, starting in the early part of the twentieth century with his grandfather, who was affectionately known as 'Paddy the Painter'. The Hiltons have always been in Sunderland; Paddy himself was born in Walworth Street, where The Bridges shopping centre now stands and not far from the spot where we meet, in a small pub called Fitzgerald's.

The venue couldn't be more appropriate: the walls are covered in pictures of ships and flourishing streets; a Sunderland that, to me, is completely unrecognisable. Growing up in the beginnings of the post-war slump for Sunderland, Paddy remembers the town going through hard times:

> A lot of the streets in the town centre were pretty desperate; they were like slums. There was a ragman called Raspy Toms who'd come round on his horse and cart, and blow a bugle for people to come out with rags and get a tuppence. The coal would come round on a horse and cart as well and got dumped in a back lane to be carried in buckets to the coal house. For a couple of bob he'd shovel it into the coal hole for you and someone would sweep up the dust and take it to burn. If you put a bit of cement in you could make bowls out of it and dry them on the fire.

Paddy, like most children living in the town centre at that time, went to St Mary's school at the bottom of Chester Road, where he excelled in woodwork and sport, but nothing academic. 'You either went down the pits or in the shipyard then, so I was destined for the shipyard. Dad went in to see the plumbing foreman and got me in as an apprentice plumber.'

I've become incredibly familiar with the phrase 'you either went to the pits or the shipyards'. Prior to interviewing anyone, I'd wondered if the expression was exaggerated, one of those working-town clichés that people trot out like a learned script at the mere hint of the 'grim north'. To the majority of men who grew up in Sunderland at that time, though, it's a simple truth.

In 1958, Paddy started an apprenticeship at William Doxford and Sons shipyard, which would go through many name changes over the years, but will always be known to the people of Sunderland as Doxford's. 'For the first few weeks you were put in the stores, so you knew which spanners and pipes were what. Then you got stuck with another guy and went on the ship and every six months you'd change the type of work you were doing.'

Every workforce needs a rascal and Paddy admits he was a handful on the job: 'I was one of the worst. If anyone was going to get told off, it would be me.

The gateway to William Doxford and Sons, Pallion, 1950s. *Sunderland Museum and Winter Gardens*

In the winter, when there were snowball fights, I let blaze with a snowball and smashed the roll of honour. I was the architect of my own downfall.' Despite his errant ways, Paddy was able to stay as an apprentice. 'When you were about 17 you got a little bag of tools and did daft jobs like curtain rails and bits of pipe between lights to pull cable through,' he explains. 'When you'd done that, they'd let you put a couple of wash basins in. If you were naughty, like I was, you got put in the double bottoms.' At this point, Paddy takes the piece of paper with my notes on, and draws a rough sketch to demonstrate. The 'double bottom' consists of the bottom two layers of the ship to form a water barrier in case the outer hull is damaged. From Paddy's drawing, it doesn't look like a pleasant place to be put to work:

> You'd never be allowed to work like that now, because it was so small, about two feet by three. You could only get a certain length of pipe in, so you had to make hundreds of them and bring them down to bolt them together, it took forever. You were there day and night seven days a week, with only a couple of candles for illumination and you could be in there for weeks so you certainly couldn't be claustrophobic. The only good thing about being down there is that nobody would bother you, so you could pile a few sacks up in the corner and have a kip!

Paddy's contempt for his bosses and the job didn't earn him any favours:

> In the summer, you were better off on the cold boat, but they'd put someone like me on fire-bending, where you had pipes that you filled with sand, heated up and you'd bend them, it was really hot work. Come the winter, when it was freezing, I'd be back on the boat.

Bartram and Sons shipyard, 1952. *Sunderland Museum and Winter Gardens*

As a shipyard apprentice in the 1950s and '60s, you were expected to supply your own work clothes until you were about 20, and then you got provided with overalls. 'They got a guy in to measure us up for them and he brought them in and nobody's fit. They had to send him back to measure up again, he'd made 100 pairs of overalls and not one pair fit.' Later, when shipyards were providing everything a worker needed, the overalls were colour coded to a particular trade. Labourers' overalls were grey and when the 'skilled' workers saw them coming, they'd make seal noises.

Like the mines and most other heavy industries, these were often dangerous places to work. 'The shipyards in the '50s and '60s should have had the same words on the gates as on cigarette packets, "Working in here can seriously damage your health",' laughs Paddy. 'You had riveters making such a noise and people banging on steel, it was so intense and painful at times.'

Health and safety concerns weren't a priority at that time and the workers relied on common sense, as well as a bit of luck:

> People would often drop things from a height: a spanner or bit of pipe, and everyone would duck and hope it didn't hit you. There was no scaffolding then so you'd put a plank across the beams and walk along, hanging on to a small beam above for dear life. The strange thing is, I can't remember anyone falling off a single plank.

The day-to-day perils of being employed in a busy industry were not to be the biggest threat to the workers' health: this came in a stealthier, more sinister form in the use of asbestos. Historically, shipbuilders are among the most affected by a condition called mesothelioma, a form of cancer linked to being exposed to asbestos, which usually starts in the lungs.

Until the 1980s, the negative effects of asbestos exposure were unknown, but the material was widely used in shipbuilding to insulate engine rooms, sleeping quarters and boiler rooms:

> You had to kneel on the iron deck to put bolts in and the pipe fitter would give you a big square of asbestos to kneel on so that you didn't hurt your knees. It was everywhere, the walls were covered, the boilers were covered – there was a sweeping-up squad, so there'd be a big cloud of dust in the air and we'd breathe it in. You used to see workers rolling the new starters in the asbestos dust so they were white from head to foot and then they'd clean them off with the compressed air. I've got pleural plaques now because of it. I got compensated, but it's no compensation.

Pleural plaques are areas of scar tissue on or around the lungs caused by exposure to asbestos. They can lead to other diseases such as asbestosis, mesothelioma or lung cancer.

My uncle, Pete Wright, had worked at the shipyards in Sunderland; this was another one of those family conversations where a significant piece of history is casually dropped in. For Pete, despite growing up in a shipbuilding town, the shipyards were something he'd ignored whenever possible. 'I didn't take much notice of it, to be absolutely honest,' he says:

> It wasn't something I was ever going to go into. All my family were in the shipyards: my uncles, cousins, my dad. I was the first in my family to pass my 11-plus and I went to Southmoor Tech, so in my mind I was never going to go in the shipyards.

Court Line had taken over Doxford and Sunderland Shipbuilding in 1972 and Pete remembers the wages increasing dramatically:

> They were a big business throughout the world and when they took over, the wages were astronomical. To give you an idea, I always remember the average wage in the whole of Britain was £45 [sources vary but most put it around £48 per week in 1974] and we were getting £75, so that's the reason I went to the shipyards. With hindsight, it was a big mistake, I never should have gone to somewhere just for the money, but at 19 we were able to buy a three-bed semi at Seaburn Dene.

Apprenticeships had changed in the fifteen years between Paddy and Pete joining the shipyards. Pete started at J.L. Thompson and Sons' North Sands yard in 1974. Unlike Paddy, who started learning with the tradesman immediately, after an initial tour of the yard Pete was sent to college for the first year. 'You were with all your mates, all the lads and it wasn't work really. We walked round, waiting to go to college, the next year we went to college for two days and the third and fourth year, we went for a day.'

Apprentices were still the butt of many jokes in Pete's day, but thankfully the days of rolling them in asbestos dust were long gone:

> With my family being in the shipyards, they warned me before I went in, don't go for a 'long stand' or the 'stripy paint', so I knew it all and when they started with the banter, I got in first. You'd always get an apprentice who didn't, though,

> and you'd try and warn them but they'd just ignore you. I never saw anything vicious, you hear of them wrapping apprentices up or hanging them from hooks and things, but I never saw anything like that. When I was in the yards, everyone got paid the same money, but at one time when a ship was being built, they'd hire loads of people and then pay them off when it was launched. So the older people were jealous of apprentices and I heard tales of them sending them off and changing the plan while they were gone, so the apprentice never learned what to do. There were a few shipyard men who couldn't read the plans, though, they'd just done the same job year after year, whereas the young lads coming through had been to college and there was a little bit of jealousy.

Being a shipwright has always been a prestigious and sought-after position at the shipyards, Pete remembers:

> After you'd been to college and got your results, they'd give you your trade. Everyone jokes about whose is the best, but out of 120 people, most of us wanted to be a shipwright or a plater. In the end there were four platers and four shipwrights; the rest were welders, caulkers or burners. Shipwrights thought they were the best, platers thought they were the best. You more or less did the same thing, but a shipwright is a carpenter outside of the yards, so you got an allowance to buy tools and things like that.

How a ship is built depends on a variety of factors. Pete remembers when his father worked in the shipyards and ships were built in sections in the shed and then moved to the ship. In his day, shipwrights were split into two teams, one building the shell of the ship and one responsible for outfitting:

> The only time we came together was when we were launching the ship. The ships at Pickersgill's [Austin and Pickersgill] were called SD14s and people always thought that meant standard design, but really it stood for sheltered deck. It was the same boat built time and time again, where at Thompson's yard, different ships were built, so you had to start from scratch every time. Draughtsmen did the plans for the ship, then those went to the loftsman who would make it in wood so you knew the shape of it and then it went down to the tradesman. The bigger ships took about five or six months in total to build.

Health and safety and working conditions in general had come on leaps and bounds between Paddy's time at the yards and Pete's:

> Court Line provided overalls, you took them off once a week and they cleaned them for you, whereas previously, you'd have to take them home and your wife would wash them. You could have ear muffs if you wanted them, but we didn't have riveters like the yards used to, it was all welding so it wasn't as noisy. When I was there, nobody ever wore a hard hat even though they were supplied to you if you wanted one. The old men wore flat caps, the young lads wore nothing. If you put a hard hat on, they thought you were soft so you just didn't do it. The things those old blokes could do with their flat caps ... they used to hold hot plates with it and everything. There was one man who was bald and he was very self-conscious. He had a cap to go home with and one to work with and at home time we'd all watch him. He'd go into the locker, take out the cap and [Pete mimes a lightning-quick cap change] everyone would cheer and laugh.

Unlike in Paddy's day, when you'd come home from work exhausted, damp and dirty, Pete saw a different, more modern side to the shipyards:

> When I was there, the plates were painted battleship grey, they weren't filthy, you didn't get dirty then so it was a total change from the early days. My generation didn't rough it down the shipyards like the older ones did. It wasn't hard work, but you were cold and damp; you'd start at 7 a.m. and it was freezing in the winter. At 9 a.m. you could have your bait for twenty minutes and we'd sit on plates the size of a house. Rats would be running around underneath while you ate. It got better while I was there, they knocked the bait places down and built a big canteen. At Pallion, they installed a coffee machine and we thought it was fantastic. We only had it for three months and they took it out when we were on our Easter holidays because they said too many people were using it! It was archaic in some ways because the buzzer would go at 9 a.m. and you'd have to get from your ship to the bait cabin, then at 9.15 the buzzer would go and you had to be back on the ship for 9.20. It was great being with the lads though, that was the bit I enjoyed.

Accidents, although infrequent, did still happen:

> On the *Aurora* [a ship built in Sunderland in the mid-1970s], lads would walk on the deck of the ship and there was a manhole with wood over, to cover it, but you never, ever stepped on it. Even to this day when I step back, I always look behind, I never just step back without looking. Well on this occasion, one man did step back and he went straight through the wood all the way to the bottom and was killed.

The launch of the MV *August Pacific*, built at the North Sands Yard, 7 March 1969. *Sunderland Museum and Winter Gardens*

Although safety wear wouldn't have prevented that particular tragedy, a number of other near-misses were still not enough to get the men wearing protective clothing. Losing face with your fellow workers played a big part in many of the lapses in health and safety; even when clothing or equipment were provided, to use them was to appear 'soft' to your workmates. Pete remembers that:

> The yards started to provide safety boots with the steel toe caps in, but mostly people just wore Dr Martens. Somebody dropped something on his foot once, and if he hadn't been wearing the safety boots, he'd have lost his foot. He got some money from the company for wearing them to encourage us all to, but still nobody else would wear them.

Thankfully Pete managed to avoid many of the dangers of the yard, but he did have a particularly hair-raising experience that has stuck in his mind: 'I once went up on to one of the massive cranes because the bogie had fallen off. I had to go up and put staging on, so they could get up and fix it.' Pete and two other workers were tasked with climbing the crane, unaided and unsecured, to put scaffolding on, enabling the engineers to get up and fix the bogie:

> The three of us just started laughing, we were so frightened, and we couldn't stop. Our feet were aching from gripping our toes inside our boots. Usually you go up in stages so you can't really see down, but on the crane you looked out and you could see everything and the people looked tiny. You had to lean over to put brackets on and then put planks on those and then step out onto the loose planks. Sometimes there was a bit of give in the brackets and you'd step on and it would go down about an eighth of an inch and you'd think your end had come, but you laughed and joked about it. I couldn't do that now, I just couldn't.

Many people think that a launched ship was a finished ship. Pete explains that there's much more to it:

> That's just the shell done. It gets taken to the dock and you have to fit it out; that's what I mostly did. So, fitting the hatches, the decking and things like that. We used to have to go underneath the ship and you called it splitting out. Where the ship was held on wooden posts all the way along, a wedge was put under and then hit to split the post. It was done in sections and finally the triggers would be gone and the ship would go down into the water.

Part of a ship's trials involves testing the equipment and the propeller. This had thrown up a particular issue for many decades: 'There were loads of times when dead bodies would come to the top. I used to see them go in and get the body, they'd get hooks and get in and then the police would come down. It was quite often, not just occasionally.' Surprisingly I'd never heard this macabre tale from the yards before, but apparently it was a common occurrence. In older times, if a body was found, there was a bounty to be paid for getting it out of the river and so men would race to get the body before anyone else. One side of the river was more lucrative than the other and if times were particularly hard, it's been documented that often, those who were exceptionally poor wouldn't always wait for a body to turn up.

The launching of a ship seems like an exciting thing, but I can imagine that if you worked with them day in and day out, it might lose its interest. I've wondered whether any ship enthusiasts worked at the yards in the same way you see train or bus spotters now. I've often caught a glimpse of the occasional bus-spotter around Sunderland and quite recently, noticed a man keenly taking a picture of a Stagecoach bus. Surely a thriving shipyard town must have inspired ship-spotters in the same way that docks or ports do now? 'Some blokes used to write down the name of the ship, how much it weighed … they were very keen. I couldn't have given a monkey's.'

No matter the background of the men, the camaraderie in the shipyards was strong. 'It was quite a tight bond once those gates closed,' Pete remembers:

> There were no females working there, a few cleaners but that was about it, so it was a totally male-oriented place. When the gates shut, the language came out but once you went back out on to the street, the men didn't swear. If someone had sworn in front of kids or women, they'd have been told to watch their language.

The older men and younger ones generally rubbed along, aside from some rare friction between some college-educated apprentices and traditionalists. Outside of the yards, though, there were notable differences, and the attitudes of the younger generation were starting to change:

> There were often funerals at St Peter's Church and the older blokes would stop and take their caps off as it went past and then put it back on again, it was a natural thing. They'd be talking, see the hearse, take their cap off and then put it back on again. That's something my generation never did.

Court Line went bust not long after Pete started his apprenticeship, and it went downhill from there. In the year that Pete joined, there were 120 people taken on as apprentices. The following year there were twelve – a stark indication of the coming decline. The government took over and the company was renamed British Shipbuilders. 'You never thought about the place going bust, you hear of it all the time now but in those days you didn't,' Pete muses:

> I realised afterwards, they could've suspended my apprenticeship but then it never crossed your mind that you might lose your job. The wages went down and down and down because of the government. If you wanted a pay rise they told you to just go if you weren't happy. At first, you could more or less pick your job and if you didn't like it you could go elsewhere. Then over the space of about four or five years you saw that stop and you couldn't pick and choose your job and there were no jobs to go to.

As shipyards and the pits started to struggle in earnest, seeing people out of work in Sunderland became a common occurrence:

> You saw more and more young lads walking around the streets with no jobs. You could really see a difference in the whole attitude. You used to get up on a morning and you'd see everyone coming to work and on a night you'd see everyone going back and the bridge was absolutely heaving. Then you stopped seeing that and started seeing people sitting around, something you never saw before; it was awful. There were loads of men who'd gone to the shipyards and bought their own houses, then they lost their jobs and homelessness went through the roof because people couldn't afford their own homes.

The scenario Pete is describing sounds sadly familiar. Unemployment is still high and you often see people hanging around with seemingly very little to do:

> I think there are people in Sunderland who haven't worked all their lives. I know some people who haven't worked now into the third and fourth generations. I don't think it's ever recovered. My friends either couldn't get another job or they went all over the world looking for work: Germany, the Netherlands, Mexico, America ... doing temporary jobs. Some went to the oil rigs but it cost a lot because you needed to pay for things like safety certificates and many couldn't afford it.

The industry, with the exception of a few smaller yards, was nationalised in 1977 as the British Shipbuilders Corporation. They had closed half of the remaining shipyards by the end of 1982 and sold off, merged or restructured many of the remaining yards. In 1983, the Conservative government, under Margaret Thatcher, announced that they were to be privatised:

> When I was young you had Monkwearmouth Colliery, the shipyards, Doxford Engine Works – ships all over the world had a Doxford engine in. All those industries where all those people used to go to work and suddenly there's nothing there. You had certain standards, you got up on a morning, you went to work and for all I didn't like working at the shipyards, I still wanted to do a good job, still took pride in it. There's none of that now.

In 1988, Thatcher declared British Shipbuilders would cease, and the last two shipyards, North East Shipbuilders Ltd and Pallion and Southwick yards (formerly Austin Pickersgill and Sunderland Shipbuilders Ltd) closed in 1989, bringing to an end hundreds of years of Sunderland's shipbuilding history.

To me, in 2017, seeing an enormous ship from the Wearmouth Bridge or at the end of your street seems like an exceptional, incredible thing. For those living on or near the river, though, this must have been an almost daily occurrence, and something that was so ordinary it wasn't worth noticing or remarking on. To put it into context, Sunderland is currently having another river crossing built, and as one of the larger pieces was shipped down the river to be put into place, crowds lined the banks – social media buzzed and news outlets filmed and reported the spectacle to an exhilarated North East – something that would've been unremarkable some thirty or forty years ago.

8

Nissan

Nissan Motor Company Ltd started in 1911, and its founder, Masujiro Hashimoto, was seen as a pioneer in Japan's automotive industry. As Nissan grew, it started to expand into foreign markets like the United States and Australia. Exporting to Europe was becoming expensive and so Nissan began to look for a suitable country in which to build a European plant. After narrowing the search to England, three sites were looked at as possibilities until finally, Sunderland was chosen.

Getting an interview with Kevin Fitzpatrick, the divisional vice president for European manufacturing at Nissan, is hard. We've not long had the EU referendum and Kevin is in demand for comments over the security of Nissan jobs in the city. Senior press officer for Nissan, Ben Guy, is the gatekeeper for Kevin; he wants to know what I'll ask, what kind of answers I'm expecting; nothing unreasonable but I get the impression tensions at the plant are high. Thankfully, soon after we make initial contact, Nissan receives some assurances from the government and also lands the contract to make a new model, securing jobs for the near future and taking a great deal of the pressure off.

Arriving at the plant, you get a sense of just how enormous it really is. Built on the former site of Sunderland Airport, in Usworth, Washington, the site sprawls as far as the eye can see, and it takes a few minutes of driving around to find out where I need to be. Once inside, I'm greeted by Ben, who is polite and pleasant and clearly already, at 8 a.m., spinning a number of plates. After a short wait I'm led upstairs to meet Kevin.

Kevin isn't what I expect. I'd pictured a London banker type, suited and booted and eager to get me out of the door as soon as possible. When Kevin walks in, he's in suit trousers with a Nissan-branded tracksuit-style jacket, carrying a cup of

Sunderland Airport, 1984. *Nissan*

coffee. I wonder whether, as I'm filming the interview, this is Ben's way of having their branding in the shot. Kevin has a local accent and I discover he grew up in Washington, which pleases me in the strange way that my admission of being from Sunderland seemed to please everyone else I interviewed – this is clearly something we all share.

Kevin looks just the right amount of uneasy to show me that this isn't his comfort zone, and despite often being called on for interviews, I'm sure he'd rather be anywhere else. It makes me appreciate the time he's taken to chat to me even more. Ben sits close by, notes in hand, ready to jump in if Kevin needs rescue and, I'm quite sure, to make sure he doesn't talk about anything he's not supposed to.

As Sunderland uses Nissan as bragging rights, I wondered how it had landed the prestigious company at a time when it was desperate for some new hope, 'It was a whole host of things really,' Kevin says:

> The key strategic reasons are the transport links we have, we've got the A19 and the A1. We have a deep sea port at Sunderland, and then we've got Newcastle airport. The other thing was the land site was very easily adapted for manufacturing cars. The plant is huge, so we needed large, flat areas and the site being an airfield fit the bill nicely. There was an availability of skills at the time and it's no secret that the government gave strong support. We [Wearside] were competing against at least another two sites in the UK. Then I think there was an emotional side to it as well. In Japan, you have the Japanese navy and a lot of the ships were built in Sunderland, so they had a good sense of our quality of work. I think it was a mix of all of these things.

Being a local, Kevin knows the area well and remembers the North East's reputation for heavy industry:

> I worked at Team Valley Trading Estate and before that I worked at the Royal Ordinance Factory – the whole of the North East was in serious decline, everything was getting smaller or closing and the traditional industries weren't doing particularly well. Caterpillar at Birtley was a huge plant and that closed, it was pretty grim in the North East, and Nissan coming here gave people something to latch on to and gave the region something to use as a showcase; if it was successful, and fortunately it was.

Nissan have around 7,000 employees at the factory in Washington and then another few thousand in the form of contractors or suppliers, so around 10,000 people come and go from the plant. I wondered if shift-change time was reminiscent of

An aerial view of Nissan prior to the 2013 extension. *Nissan*

the famous photographs and footage of the Sunderland shipyards, where the gates open and a sea of workers stream out:

> They do that, but they do it in cars now, not on their feet and not in overalls. It's difficult to manage, we've got two production lines and we stagger the start and finish times by a few minutes to ease it, but when the day shift's working overtime and the late shift's coming, we end up with two full crews in at the same time, so we need to double the car parking space for changeovers. We were caught out on planning with the recent expansion we've had. We thought we had enough car parking space and we'd done all the surveys, but then when you get into the practicality of changing shifts, it didn't work and we've had to extend the car parks.

When jobs in heavy industry decline, there are a lot of skilled people out of work and so the businesses who are still thriving can take their pick of workers. Eventually, though, those workers reach retirement age and the next generation aren't trained or skilled in the same jobs. This is something Nissan has experienced first-hand while in Sunderland. 'One thing we do a lot of came from the skill shortages we started experiencing about four or five years ago,' explains Kevin:

> There weren't enough graduates, maintenance technicians, tool makers and so on, in industry, not just automotive but manufacturing in general. Enough people weren't being trained because it's very expensive. When industry is declining, you don't have a skill shortage because there are plenty of people coming on to the labour market that you can use. But about five years ago, we seemed to hit the bottom in terms of places closing; we started expanding and the skills just weren't there, so the remaining industries all started taking each other's people and we thought, we need to start training. We noticed that there weren't enough kids in schools taking STEM [science, technology, engineering and mathematics] subjects, so we started engaging with schools and teaching the art of making things. We have programmes where we take a bus to school and teach them the concepts of manufacturing, in a fun way; we do it with Lego and they have races. We also teach them about electric cars, sustainability and renewable energy. We bring 2,000 school kids a year here, to the plant and we do F1 for schools, where a former racing team design and make small cars, and market them.

I'd never given much thought to the skills left to wane and die when the shipyards and mines went. At one time, it was expected that men would know how to use their hands to craft, to make, to build or to dig. So how do you entice an up-and-coming generation back into the types of jobs their great-grandparents likely had? 'Schools don't have the resources to advise parents and the children on careers, you don't have traditional careers masters any more,' observes Kevin. I think back to my school days and I vaguely remember having a careers test that consisted of filling in a questionnaire and it being sent away for a computer to determine your top three careers choices. I can remember many of the kids being inconsolable, thinking this meant their dreams of becoming an astronaut or a footballer were out of the window. I can't remember what my test suggested to me, but I'm fairly sure that the jobs I've had since weren't on the list.

Kevin continues:

> Parents have watched manufacturing decline over the years so they think it's a bad place to work. They think it's dirty, but actually you can't build high-quality manufacturing goods in dirty premises, so factories tend to be very clean and working conditions very good. It's all about trying to reverse those perceptions and to demonstrate that actually, the jobs are quite well paid too. A manufacturing job is probably higher paid than the equivalent job in services. The gateway for progression is much better as well; people get certain types of degrees and then can't find a job because employers aren't interested. With an engineering degree, there are very few graduates in my experience who don't have choices of jobs once they graduate. I think we're starting to turn a corner now, the government are pushing apprenticeships as being good and big employers are leading the way in that regard. We have several apprenticeship programs: about 170 Level 3 apprentices, 200 manufacturing staff and over 100 graduates. We like to use it as a gateway into the company, as we don't actually recruit much from outside.

It seems like a smart move to educate, train and guide from the start, getting kids excited about jobs like engineering again. Around the time I was leaving school, there was a big push from the government to encourage people to become teachers, including cash incentives for taking education related courses. Many of my friends did this and very few are now still teaching; the field became saturated with individuals who had similar experience, qualifications and skills.

I remember at school that an IT teacher, for example, might be off sick and you'd be taught by your French teacher who knew nothing about computers. In the days of heavy industry, as we heard from Pete who worked at the shipyards, the men could do a variety of jobs to make sure that business could carry on as usual,

Ground breaking at Nissan, 1984. *Nissan*

if something were to happen. Nissan have a similar approach: 'We call it three jobs one man, one man three jobs,' says Kevin:

> It's quite difficult on an ongoing basis because people move to other areas, people leave or new starters come in. We try to have it so that three jobs can be done by one man and one job has three men that can do it; normally the best we do is about two and a half. The absolute minimum is two, we can't run properly unless every guy can do two jobs and each job has two men.

The local universities have been a major part of helping to populate Nissan with educated, skilled workers:

> We have links with the universities and take about sixty graduates a year. In the North East there are about 60–70,000 university students, so we don't put a tiny scratch in that amount of people, but what we tend to do, is work with them on placements at our development site at Cranfield where we've got our Design Centre.

Workers in industry generally followed in their parents' footsteps. Albert the miner, forced to work at Monkwearmouth Colliery for fear of his father losing his job and their home; Pete, who despite different circumstances entirely, ended up working in the shipyards for a time just like his father. Times have moved on still further, so now, with new industry bringing better opportunities and a cleaner and safer work environment, are we starting to see a new era of an inherited workforce?

> We have three generations, it's not common but we do. Two generations is pretty normal; my son works here for example. It's only in the last five or six years that you start to notice 'oh that's so-and-so's son'. It's not like the old days when you left school, because you've got more choices now, but it's quite good, I think it gives the place more of an atmosphere.

Unlike the shipyards and collieries, which were, for the most part, male-oriented, today we live in a more progressive society. Schemes are being rolled out left, right and centre to entice young women into jobs that were traditionally thought of as men's roles, but is it working? 'We don't get a lot interest from females in manufacturing jobs, but in the offices and even in engineering it is increasing. We deliberately try to get more female engineers because, as a ratio, they get further than male engineers, but there just aren't the numbers going through university.'

With Sunderland voting to leave the EU, I was keen to find out whether the new-found fear for Nissan's long-term survival was warranted. I grin sheepishly at Ben as I ask the question, because I know this is the kind of topic that warrants a carefully crafted answer; not because it would require deception but because the reality is that there probably isn't a definitive answer at this early stage, and any reply given, must be measured, considered and not able to bite you on the behind at a later date.

> I think the relationship we've got with Sunderland council is great, very constructive and they're always there if we need help. To Sunderland, Nissan is a big employer and big employers create jobs, and we're one of the types of employer where when we win a new car, you can see forward five years; there aren't many industries or companies where you've got that kind of stability. If you're not winning new models, obviously that would be a different story. One thing that isn't good, is Sunderland is too reliant on Nissan because there aren't enough big employers. When I was in my early 20s you could

reel them off and they've all gone. The uncertainty that faces the country at the moment isn't a good thing, but there's nothing we can do about it because the UK has voted to come out of the EU. Nissan has very publicly made a commitment, which has taken away some of the uncertainty from us, but not from the country, so we just have to wait and see what happens and hope that the government have a very ordered and pragmatic approach where everyone wins.

Conclusion

What lies in Sunderland's future? With the closures of the major industries still within living memory, stories are passed down the generations. This is an integral part of the way we learn from and understand our history, and what better representation than to be told by the people who were there?

Nissan has been brought up in conversation repeatedly, it appears in this book several times, mentioned by interviewees in glowing, almost reverent terms: 'Nissan saved Sunderland', 'We'd be lost without them', but Nissan's Kevin Fitzpatrick warns that Sunderland is too reliant on Nissan, and indeed, it seems that Nissan is a cast, doing its best to keep a fractured Sunderland together, but perhaps not healing it. In the wake of the Brexit vote, doubt has been cast upon the longevity of Nissan in the city and although some fears have been temporarily allayed, the long-term future is uncertain.

I'm reminded of the Monty Python scene, 'What have the Romans ever done for us?' What has the EU done for the North East of England and Sunderland, specifically? The European Regional Development Fund (ERDF) was created to help balance regions within the EU, to attempt to get everyone on a more even footing. Sunderland Software City, was built with around £5 million from this fund; Washington Business Centre received almost £3.5 million, and the University of Sunderland got £1.3 million, to help new graduates find work in local businesses.

Sunderland's jewel in the crown is its clean beaches, Seaburn and Roker – both of which have been awarded blue flag status on a number of occasions. What many may not be aware of, though, is that the EU's 'bathing water directives' have played a major role in this. In 1990, both Roker and Seaburn (as well as a number of other North East beaches) were classified as too dirty to swim in.

News reports shortly after the vote presented Sunderland as the poster child for Brexit, in part because the Sunderland vote came in first and showed a clear majority wanted out of the EU. In the days that followed, the images and interviews chosen to demonstrate this were not necessarily representative of the city as a whole, and painted it in that same light I touched on in the introduction. It's understandable that as Sunderland's identity is created for it outside of the city, it only serves to fuel that sense of disappointment, anger and abandonment.

Many Mackems see the town that was strong, affluent and industrious and compare it to what they have today: queues at the job centre, shops and factories standing vacant and vast, empty brownfield sites with unfulfilled potential. It's easy to forget that circumstances were different – the two world wars created a need for goods and services outside of the norm. Shipbuilding and mining, in particular, were subsidised, and as practical and environmental factors changed, that was no longer worthwhile.

When central government pulled the rug and stopped subsidising major industries, it was inevitable that a place like Sunderland, highly reliant on those workplaces, would suffer greatly. The wonder is that it would gaze longingly back

Liebherr, Ayres Quay is coming up to its twenty-eighth year in Sunderland. *Marie Gardiner*

to that time, while in tandem voting essentially to end any support received from EU funding.

Despite this habit of looking backwards, Sunderland has, albeit slowly, been trying to move forwards. The mid-1980s saw a huge regeneration; as we know, Nissan arrived at a time when Sunderland needed it most. Doxford International Business Park was created in 1990 on a former greenfield site, designed to encourage growth and development in response to the disappearance of the shipyards and mines.

The University of Sunderland today is unrecognisable from its modest start in 1901 as Sunderland Technical College and School of Art. In the 1920s it offered, perhaps unsurprisingly, a Naval Architecture course and it was also the first to offer courses around the making of sandwiches – something like our modern food preparation courses.

A mining department was added in 1930, and during the Second World War the college ran special courses for the armed forces. In 1969, the Technical College, the School of Art and Sunderland Teacher Training College merged to form Sunderland Polytechnic. In 1992 it gained university status and currently has three campuses, one in London and two in Sunderland; Chester Road and St Peter's, the site of the monastery built by Biscop in the seventh century.

The university is highly commended, and has been shortlisted for a number of awards, including University of the Year; it won the *Times* award for Best Student Experience in 2005. Some 13,000 students attend the university, contributing to Sunderland's economy during the time they're there.

Reputable companies such as Berghaus and Arriva have their headquarters in Sunderland, the latter actually being a home-grown Sunderland company founded by Tom Cowie before becoming part of German Deutsche Bahn. Liebherr arrived in Sunderland twenty-seven years ago and helps to retain some of the proud maritime tradition with its cranes and cargo equipment.

The Sunniside area of the city centre was redeveloped in 2004 to include a cinema, restaurants, bowling alley, casino and multi-storey car park. Although the idea was for Sunniside Leisure (as it's now known) to become the cultural centre of Sunderland, that didn't really happen and now the idea of that hub has shifted to the new developments at Keel Square and the old Vaux site.

Building on the sites where the major industries used to stand is a positive and cathartic thing. Where Monkwearmouth Colliery stood, we now have Stadium Village, home to the Stadium of Light, Sunderland Aquatic Centre, the Hilton Garden Inn and soon, the Beacon of Light; all bringing tourism, jobs and money to the city. The National Glass Centre and the University of Sunderland are on the sites of the old monastery, the very beginnings of glass manufacturing in Sunderland, and the former home of a major shipyard. Frank Nicholson described the empty

Vaux site as a physical and emotional scar to Sunderland and now that plans are finally in place to build something there it seems to have instilled a new positivity in the city; the people of Sunderland have something to look forward to again.

The new Wear Crossing, currently being built, has generated much excitement. Although a practical necessity to relieve traffic congestion and strengthen the transport links by joining the A1231 with Pallion, it's also become a symbol, heralding the next stage of Sunderland's evolution.

This move from heavy industry to a focus on education, art, heritage and culture is highlighted by the fact that at the time of writing, Sunderland is bidding for the City of Culture. The bid has energised the city, galvanising conversations and spurring action and new ideas. One thing Sunderland has been guilty of in the past is looking at what others have done and buying 'off the shelf' events in the hope of replicating success. The City of Culture bid has forced the council and associated enterprises to start thinking innovatively, and to make use of the great resources at their feet and trust in their own people, use their own talent, like local musicians, artists and historians.

The Arts Council has given the newly formed MAC (media, arts and culture) Trust a £6 million injection to create a brand-new venue to host some of the city's new events. The old fire station is being revamped at last, after plans were put on hold over two years ago – it will become a community space to hold classes and activities.

It seems that, finally, Sunderland is coming out of its own shadow. We must, as people and as a city, be prepared to move forwards. There's a danger that too much harking back can mean that we give up on our future. When we're looking at what we had and what we've lost, it's important to remember what we still have, to celebrate our successes and to ensure that the next generation of social histories is just as interesting, fulfilling and hopeful as the last.

What Sunderland needs, and is moving towards, is a new identity; one not created for it by those outside looking in, and not by looking back at what it used to be. It needs an identity that is created by the people who live and work there. Something to feel fulfilled by, something that, when someone asks, makes us proud to look them in the eye and say yes, I belong to Sunderland.

Further Reading

Online Resources

BBC Tyne and Wear: www.bbc.co.uk
Bob Murray: www.sirbobmurray.com
The Co-op: www.co-operative.coop
The Chronicle: www.chroniclelive.co.uk
Durham in Time: www.durhamintime.org.uk
Durham Mining Museum: www.dmm.org.uk
Engineering Timelines: www.engineering-timelines.com
England's North East: www.englandsnortheast.co.uk
Grace's Guide to British Industrial History: www.gracesguide.co.uk
Institute of Historical Research: www.history.ac.uk
ITV Tyne Tees: www.itv.com
The Journal: www.thejournal.co.uk
The National Glass Centre: www.nationalglasscentre.com
Newcastle University: www.ncl.ac.uk
The Northern Echo: www.thenorthernecho.co.uk
Roker Pier: www.rokerpier.co.uk
Searle Canada: www.searlecanada.org
Sunderland Antiquarian Society: www.sunderland-antiquarians.org
Sunderland City Council: www.sunderland.gov.uk
The Sunderland Echo: www.sunderlandecho.com
Sunderland Maritime Heritage: www.sunderlandmaritimeheritage.org.uk
Tyne and Wear's Historic Environment Record: www.twsitelines.info
Victoria County History: www.victoriacountyhistory.ac.uk

Books

Curtis, J., *The Engineer and the MP* (Newcastle: Summerhill Books, 2014)
Dodds, G.L., *A History of Sunderland* (Surrey: Albion Press, 2001)
Nicholson, P., *Brewer at Bay* (Durham: Memoir Club, 2013)

If you enjoyed this book, you may also be interested in…

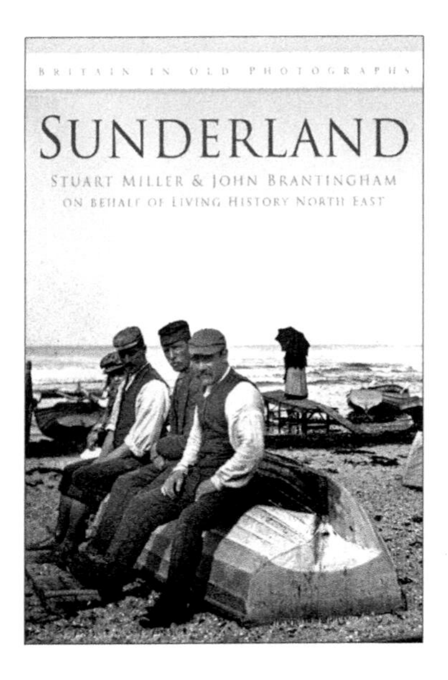

Sunderland

STUART MILLER & JOHN BRANTINGHAM

978 0 7524492 7 2

This collection is comprised almost entirely of images which have never previously appeared in print. Starting in the earliest days of photography, travelling through the war years to the post-war period and finishing with the shipyards and pits of the 1970s, in the last days of those industries, this collection will delight and amaze in equal measure.

This volume draws upon the extensive research of Living History North East. With the memories of Sunderland residents long gone and the contribution of many of the area's current residents, it is a book that celebrates every aspect of life in the area. It will enthral residents and visitors alike.